COMPUTATIONAL OPTIMIZATION, MODELING, AND SIMULATION FOR ENGINEERING APPLICATIONS

AAP Research Notes on Optimization and Decision-Making Theories

COMPUTATIONAL OPTIMIZATION, MODELING, AND SIMULATION FOR ENGINEERING APPLICATIONS

Edited by

Anupam Shukla, PhD
Sourabh Rungta, PhD
Mohan Awasthy, PhD
Rakesh L. Himte, PhD

APPLE ACADEMIC PRESS

First edition published 2025

Apple Academic Press Inc.
1265 Goldenrod Circle, NE,
Palm Bay, FL 32905 USA

760 Laurentian Drive, Unit 19,
Burlington, ON L7N 0A4, CANADA

CRC Press
2385 NW Executive Center Drive,
Suite 320, Boca Raton FL 33431

4 Park Square, Milton Park,
Abingdon, Oxon, OX14 4RN UK

Library and Archives Canada Cataloguing in Publication

CIP data on file with Canada Library and Archives

Library of Congress Cataloging-in-Publication Data

CIP data on file with US Library of Congress

ISBN: 978-1-77491-600-1 (hbk)
ISBN: 978-1-77491-601-8 (pbk)
ISBN: 978-1-00345-402-1 (ebk)

ABOUT THE AAP RESEARCH NOTES ON OPTIMIZATION AND DECISION-MAKING THEORIES

Most real-world search and optimization problems naturally involve multiple criteria as objectives. Different solutions may produce trade-offs (conflicting scenarios) among different objectives. A solution that is better with respect to one objective may be a compromising one for other objectives. This compels one to choose a solution that is optimal with respect to only one objective. Due to such constraints, multi-objective optimization problems (MOPs) are difficult to solve since the objectives usually conflict with each other. It is usually hard to find an optimal solution that satisfies all objectives from the mathematical point of view. In addition, it is quite common that the criteria of real-world MOPs encompass uncertain information, which becomes quite a challenging task for a decision maker to select the criteria. Also, the complexities involved in designing mathematical models increase. Considering, planning, and appropriate decision-making require the use of analytical methods that examine trade-offs; consider multiple scientific, political, economic, ecological, and social dimensions; and reduce possible conflicts in an optimizing framework. Among all these, real-world multi-criteria decision-making (MCDM) problems related to engineering optimizations are categorically important and are quite often encountered with a wide range of applicability.

MCDM problems are basically a fundamental issue in various fields, including applied mathematics, computer science, engineering, management, and operations research. MCDM models provide a useful way for modeling various real-world problems and are extensively used in many different types of systems, including, but not limited to, communications, mechanics, electronics, manufacturing, business management, logistics, supply chain, energy, urban development, waste management, and so forth.

In the aforementioned cases, modeling of multiple criteria problems often becomes more complex if the associated parameters are uncertain and imprecise in nature. Impreciseness or uncertainty exists within the parameters due to imperfect knowledge of information, measurement uncertainty, sampling uncertainty, mathematical modeling uncertainty, etc. Theories like probability theory, fuzzy set theory, type-2 fuzzy set theory, rough set, grey theory, neutrosophic uncertainty theory available in the existing literature deal with such uncertainties. Nevertheless, the uncertain multi-criteria characteristics in such problems are not explored in depth, and a lot can be achieved in this direction. Hence, different mathematical models of real-life multi-criteria optimization problems can be developed on various uncertain frameworks with special emphasis on sustainability, manufacturing, communications, biomedical, electronics, materials, energy, agriculture, environmental engineering, strategic management, flood risk management, supply chain, waste management, transportations, economics, and industrial engineering problems, to name a few.

This series reports on current trends and advances in optimization and decision-making theories in a wider range of domains for academic and research institutes along with industrial organizations.

Features of the volumes include recent trends, model extensions, developments, real-time examples, case studies, and applications. The volumes aim to serve as valuable resources for undergraduate, postgraduate, and doctoral students, as well as for researchers and professionals working in a wider range of areas.

CURRENT & FORTHCOMING BOOKS IN THE SERIES

Multi-Criteria Decision-Making Techniques in Waste Management: A Case Study of India
Editors: Suchismita Satapathy, Debesh Mishra, and Prasanjit Chatterje

Applications of Artificial Intelligence in Business and Finance: Modern Trends
Editors: Vikas Garg, Shalini Aggarwal, Pooja Tiwari, and Prasenjit Chatterjee

Advanced Computer Science Applications: Recent Trends in AI, Machine Learning, and Network Security
Editors: Karan Singh, Latha Banda, and Manisha Manjul

Decision-Making Models and Applications in Manufacturing Environments
Editors: Pushpdant Jain, Kumar Abhishek, and Prasenjit Chatterjee

Precision Agriculture for Sustainability: Use of Smart Sensors, Actuators, and Decision Support Systems
Editors: Narendra Khatri, Ajay Kumar Vyas, Celestine Iwendi, and Prasenjit Chatterjee

Attacks on Artificial Intelligence: The New Facets of Cyber Ecospace
Editors: Kukatlapalli Pradeep Kumar, Boppuru Rudra, Vinay Jha Pillai, and Prasenjit Chatterjee

Computational Optimization, Modeling, and Simulation for Engineering Applications
Editors: Anupam Shukla, Sourabh Rungta, Mohan Awasthy, and Rakesh L. Himte

Advances in Intelligent Systems: Paradigms and Applications
Editors: Amit Krishna Dwivedi, Manisha Guduri, Rajanikanth Aluvalu, and Uma Maheshwari V.

Advances in Data Science and Computing Technology: Methodology and Applications
Editors: Suman Ghosal, Amitava Choudhury, Vikram Kumar Saxena, Arindam Biswas, and Prasenjit Chatterjee

Multimedia Security: Tools, Techniques, and Applications
Editors: Bhaskar Mondal and Shyam Singh Rajput

Sustainable Development: A Geospatial and Statistical Perspective
Editors: Qazi Mazhar Ali, PhD, Rizwan Ahmad, PhD, Irfan Ali, PhD, and Prasenjit Chatterjee, PhD

Advances in Intelligent Systems
Editors: Manisha Guduri, PhD, Uma Maheswari V, PhD, Rajanikanth Aluvalu, PhD, and Amit Krishna Dwivedi, PhD

ABOUT THE EDITORS

Anupam Shukla, PhD
Director, Indian Institute of Information Technology (IIIT), Pune, Maharashtra, India

Anupam Shukla, PhD, is currently serving as the Director of the Indian Institute of Information Technology (IIIT), Pune. He is an internationally renowned academician, researcher, and author. Prior to joining this institute, he was working as a Professor in the Department of Information and Communication Technology (ICT) at ABV-Indian Institute of Information Technology and Management (ABV-IIITM), Gwalior.

He has around 32 years of teaching, research, and administrative experience. He is globally renowned for his research in the field of artificial intelligence (AI), which has won him several academic accolades and resulted in collaborations with academicians across the world.

He has received several awards for his work, including "Distinguished Professor" by the Computer Society of India, Mumbai in 2017; "Dewang Mehta National Education Award" for best professor, 2016; "Prof. Rais Ahmed Memorial Award, Acoustical Society of India," in 2015; "Engineering Professor of the Month," Higher Education Review, India, 2014; "Best Paper Award," Institute of Engineers, India, 2006; University Gold Medal (Valedictorian) from Jadavpur University, Kolkata, 1998, for scoring highest marks at master's level; and "Young Scientist Award," Madhya Pradesh Council of Science and Technology, Bhopal, 1995.

He is the author of four patents, four books published by international publishers such as CRC Press, USA and Springer-Verlag, Berlin; and 190 peer-reviewed publications. He has mentored of 18 doctorate and 116 postgraduate theses and has successfully completed 13 government-sponsored projects aimed at developing IT applications for Indian farmers, skill development of the Indian youth, infrastructure development at parent institutes, etc.

At ABV-IIITM, Gwalior, he was a member of the Board of Governors for five years and a member of the senate for 12 years. He was also a member of the HR committee and the Chairman of the Student Welfare Society at ABV-IIITM, Gwalior. At NIT Raipur, he was the founding Head

Sourabh Rungta, PhD

Lecturer/Professor, Rungta College of Engineering & Technology, Bhilai, India

Sourabh Rungta, PhD, is working as a Lecturer/Professor at the Rungta College of Engineering & Technology, Bhilai, India. Previously, he served as the Dean of Computer Science and Information Technology at Chhattisgarh Swami Vivekanand Technical University, also in Bhilai, India. Dr. Rungta has published papers in various journals and conferences. He is currently supervising PhD students and has filed for one patent. Dr. Rungta holds memberships in several professional organizations, including the British Computer Society, Association of Computer Machinery, and American Society for Engineering Education. He is also a senior member of IEEE and a life member of IE Calcutta (The Institute of Engineers), ISTE Delhi (Indian Society of Technical Education), and CSI Mumbai (Computer Society of India). Dr. Rungta completed his BE in Computer Technology at Nagpur University, Nagpur, India. He obtained his MTech in Computer Technology from Pt. Ravi Shankar Shukla University, Raipur, and his PhD from CG Swami Vivekananda University, Bhilai, India.

Mohan Awasthy, PhD

Principal, G. H. Raisoni Institute of Engineering and Technology, Nagpur, India

Mohan Awasthy, PhD, serves as the Principal of the G. H. Raisoni Institute of Engineering and Technology in Nagpur, India. He has authored research papers that have been published in both national and international journals and conferences. In 2006, he was honored with a best paper award at the 21st National Convention of the Institution of Engineers. Additionally, he holds a patent under his name. Currently, he supervises two PhD students. He is a member of various professional organizations, including IEEE, Institution of Engineers (India), Indian Society for Technical Education, and Association of Computing Machinery, among others. He completed his BE in Electronics from Nagpur University, MTech in Computer Technology from NIT Raipur, and PhD in Electronics and Telecommunication from Chhattisgarh Swami Vivekanand Technical University in Bhilai, India.

Rakesh L. Himte, PhD
Professor and Head, Mechanical Engineering and Dean (IQAC), Rungta College of Engineering and Technology, Bhilai, Chhattisgarh, India

Rakesh L. Himte, PhD, is a Professor and Head of Mechanical Engineering and Dean (IQAC) at Rungta College of Engineering & Technology in Bhilai, India. He has over 22 years of teaching experience and is a member of several professional societies, including the International Society for Technology in Education, the Society for Materials Chemistry, and the Association for Machines and Mechanism. He has published papers in national and international conferences and journals. He is currently supervising several MTech projects and PhD scholars. He obtained a BE degree in Production Engineering from Amravati University, while his MTech and PhD degrees were obtained from Nagpur University in Production Engineering and Mechanical Engineering, respectively.

CONTENTS

CONTRIBUTORS

Ajay Vikram Ahirwar
Assistant Professor, Department of Civil Engineering, National Institute of Technology, Raipur, Chhattisgarh, India

Sameer Ahmed
UG Student, Electrical Engineering, RCET, Bhilai, Chhattisgarh, India

Tejasvi Anant
Department of Mechanical Engineering, Rungta College of Engineering and Technology, Bhilai, Chhattisgarh, India

N. Arjun
Associate Professor, Malla Reddy Institute of Engineering and Technology, Hyderabad, Telangana, India

Sapna Arya
UG Student, Electrical Engineering, RCET, Bhilai, Chhattisgarh, India

Mohan Awasthy
Principal and Professor, Rungta College of Engineering and Technology, Bhilai, Chhattisgarh, India

J. Ebenesar Anna Bagyam
Department of Mathematics, Avinashilingam Institute for Home Science and Higher Education for Women, Coimbatore, Tamil Nadu, India

Suraj Bandhekar
Department of Mechanical Engineering, Rungta College of Engineering and Technology, Bhilai, Chhattisgarh, India

Moon Banerjee
Associate Professor, Koneru Lakshmaiah Educational Foundation, Deemed to be University, Hyderabad, Telangana, India

Tanmay Barge
Department of Mechanical Engineering, Vishwakarma Institute of Information Technology, Pune, Maharashtra, India

Shaik Iqbal Ali Basha
Department of Mechanical Engineering, Rungta College of Engineering and Technology, Bhilai, Chhattisgarh, India

Sandeep D. Bhad
Department of Electronics and Telecommunication Engineering, Rungta College of Engineering and Technology, Bhilai, Chhattisgarh, India

Shubham Bhojane
Department of Mechanical Engineering, Vishwakarma Institute of Information Technology, Pune, Maharashtra, India

Vivek Borkar
Mechanical Engineering Department, Rungta College of Engineering and Technology, Bhilai, Chhattisgarh, India

Manoj Kumar Chande
Department of Applied Mathematics, Shri Shankaracharya Institute of Professional Management and Technology, Raipur, Chhattisgarh, India

Rahul Chowdhury
Mechanical Engineering Department, Rungta College of Engineering and Technology, Bhilai, Chhattisgarh, India

Sachin M. Dandage
Assistant Professor, Department of Computer Science and Engineering, Pankaj Laddhad Institute of Technology and Managements Studies, Buldhana, Maharashtra, India

Jayesh Deore
Department of Mechanical Engineering, Vishwakarma Institute of Information Technology, Pune, Maharashtra, India

Shrutika A. Deshmukh
Research Scholar, 2nd Year M.E. in Department of Computer Science and Engineering, Pankaj Laddhad Institute of Technology and Managements Studies, Buldhana, Maharashtra, India

Kunal Dewangan
Mechanical Engineering, Kalinga University, Naya Raipur, Chhattisgarh, India

Peeyush Dewangan
Department of Mechanical Engineering, Rungta College of Engineering and Technology, Bhilai, Chhattisgarh, India

Ritesh Kumar Dewangan
Department of Mechanical Engineering, Rungta College of Engineering and Technology, Raipur, Chhattisgarh, India

Satyendra Kumar Dewangan
Research Scholar, Department of Civil Engineering, National Institute of Technology, Raipur, Chhattisgarh, India

R. S. Dharani
Department of Mathematics, Avinashilingam Institute for Home Science and Higher Education for Women, Coimbatore, Tamil Nadu, India

Ambarish Ghosh
Research Scholar, CSVTU, Bhilai, Chhattisgarh, India

Ajay Kumar Gupta
Department of Pharmacy, Guru Ghasidas Vishwavidyalaya (A Central University), Bilaspur, Chhattisgarh, India

B. G. V. S. Jagan
Department of Pharmacy, Guru Ghasidas Vishwavidyalaya (A Central University), Bilaspur, Chhattisgarh, India

Sanmati K. Jain
Department of Pharmacy, Guru Ghasidas Vishwavidyalaya (A Central University), Bilaspur, Chhattisgarh, India

P. M. Jawandhiya
Principal, Pankaj Laddhad Institute of Technology and Managements Studies, Buldhana, Maharashtra, India

Chanchal Deep Kaur
Professor, Rungta College of Pharmaceutical Science and Research, Raipur, Chhattisgarh, India

Parul Khatri
Department of Computer Science and Engineering, OIST, Bhopal, Madhya Pradesh, India

Pradeep Kumar
Department of Electronics and Telecommunication Engineering, Rungta College of Engineering and Technology, Bhilai, Chhattisgarh, India

Sandeep Kumar
UG Student, Electrical Engineering, RCET, Bhilai, Chhattisgarh, India

Suraj Kushwaha
Department of Computer Science and Engineering, OIST, Bhopal, Madhya Pradesh, India

C. Beulah Christalin Latha
Department of Digital Sciences, Karunya Institute of Technology and Sciences, Coimbatore, Tamil Nadu, India

Mahendra Kumar Malviya
Department of Mechanical Engineering, Government Polytechnic College, Betul, Madhya Pradesh, India

Shadab Manshuri
UG Student, Electrical Engineering, RCET, Bhilai, Chhattisgarh, India

Dilip Mishra
Faculty of Science and Technology, ICFAI University, Raipur, Chhattisgarh, India

Pankaj Kumar Mishra
Department of Electronics and Telecommunication Engineering, Rungta College of Engineering and Technology, Bhilai, Chhattisgarh, India

Nikhil Nagdev
Department of Mechanical Engineering, Vishwakarma Institute of Information Technology, Pune, Maharashtra, India

Parveen Nehra
Assistant Professor, Rungta College of Pharmaceutical Science and Research, Raipur, Chhattisgarh, India

Gulab Pamnani
Mechanical Engineering Department, Malaviya National Institute of Technology Jaipur (MNIT), Rajasthan, India

Rityuj Singh Parihar
Mechanical Engineering, Shri Shankaracharya Institute of Professional Management and Technology, Raipur, Chhattisgarh, India

Hargobind Patel
Department of Computer Science and Engineering, OIST, Bhopal, Madhya Pradesh, India

Preeti Patel
ISF College of Pharmacy, Moga, Punjab, India

Vijay K. Patel
Dr. C. V. Raman Institute of Pharmacy, Dr. C. V. Raman University, Kota, Bilaspur, Chhattisgarh, India

Tejas Patil
Department of Mechanical Engineering, Vishwakarma Institute of Information Technology, Pune, Maharashtra, India

M. K. Pradhan
Department of Mechanical Engineering, MANIT, Bhopal, Madhya Pradesh, India

Harish Rajak
Department of Pharmacy, Guru Ghasidas University, Bilaspur, Chhattisgarh, India

Kantilal Rane
Professor, Koneru Lakshmaiah Educational Foundation, Deemed to be University, Hyderabad, Telangana, India

Kasi Raja Rao
Mechanical Engineering Department, Rungta College of Engineering and Technology, Bhilai, Chhattisgarh, India

Ram Krishna Rathore
Department of Mechanical Engineering, Rungta College of Engineering and Technology, Bhilai, Chhattisgarh, India

Divya Sahu
Rungta College of Pharmaceutical Science and Research, Raipur, Chhattisgarh, India

Gita Sahu
Assistant Professor, Shri Rawatpura Sarkar University, Raipur, Chhattisgarh, India

Mohit Kumar Sahu
Mechanical Engineering Department, ITM University, Raipur, Chhattisgarh, India

Sagar Sahu
Columbia Institute of Pharmacy, Raipur, Chhattisgarh, India

Roshni Sharma
Faculty of Pharmacy, Rungta College of Pharmaceutical Science and Research, Raipur, Chhattisgarh, India

Ashok Kumar Shukla
Department of Mechanical Engineering, Rungta College of Engineering and Technology, Bhilai, Chhattisgarh, India

Nitish Kumar Singh
School of Mechanical Engineering, VIT, Bhopal University, Bhopal, Madhya Pradesh, India

Ritika Singh
Assistant Professor, Rungta College of Pharmaceutical Science and Research, Raipur, Chhattisgarh, India

Shreya Singh
Faculty of Pharmacy, Rungta College of Pharmaceutical Science and Research, Raipur, Chhattisgarh, India

Agnivesh Kumar Sinha
Mechanical Engineering Department, Rungta College of Engineering and Technology, Bhilai, Chhattisgarh, India

Avinash Somatkar
Department of Mechanical Engineering, Vishwakarma Institute of Information Technology, Pune, Maharashtra, India

Vandana Somkunwar
Department of Mechanical Engineering, NITTTR, Bhopal, Madhya Pradesh, India

Vinay Kumar Soni
Mechanical Engineering Department, Columbia Institute of Engineering and Technology, Raipur, Chhattisgarh, India

Vinay Soni
Mechanical Engineering Department, Columbia Institute of Engineering and Technology, Raipur, Chhattisgarh, India

M. Sreekanth
Assistant Professor, Malla Reddy Institute of Engineering and Technology, Hyderabad, Telangana, India

Vijendra Suryavanshi
Assistant Professor, MJ College of Pharmacy, Bhilai, Chhattisgarh, India

Balram Damodhar Timande
Department of Electronics and Telecommunication Engineering, Rungta College of Engineering and Technology, Bhilai, Chhattisgarh, India

Nitin Upadhyay
Mechanical Engineering Department, Madhav Institute of Technology and Science, Gwalior, Madhya Pradesh, India

Snehal Vairagade
Department of Computer Science and Engineering, MATS University, Raipur, Chhattisgarh, India

Albert John Varghese
Assistant Professor, Electrical Engineering, RCET, Bhilai, Chhattisgarh, India

B. Vennila
Department of Mathematics, Avinashilingam Institute for Home Science and Higher Education for Women, Coimbatore, Tamil Nadu, India

Manoj Verghese
Professor, RCET, Bhilai, Chhattisgarh, India

Pravin Verma
Assistant Professor, Shri Rawatpura Sarkar University, Raipur, Chhattisgarh, India

Rinkee Verma
Assistant Professor, Rungta College of Pharmaceutical Science and Research, Raipur, Chhattisgarh, India

A. Viswanathan
Department of Mathematics, SNS College of Technology, Coimbatore, Tamil Nadu, India

J. H. Vyas
Professor, RIT, Raipur, Chhattisgarh, India

Ramesh Kumar Yadav
Faculty of Science and Technology, ICFAI University, Raipur, Chhattisgarh, India

Shrikant Zade
Faculty, Department of Computer Science and Engineering, OIST, Bhopal, Madhya Pradesh, India

ABBREVIATIONS

AD	Alzheimer's disease
ADEM	acute disseminated encephalomyelitis
ADMET	absorption, distribution, metabolism, excretion, and toxicity
AFM	atomic force microscopy
AI	artificial intelligence
Al2O3	aluminum oxide
ALK	anaplastic lymphoma kinase
AMC	aluminum metal composites
API	active pharmaceutical ingredient
AR	androgen receptor
ATDA	2-amino-1,3,4-thiadiazole
BBB	blood-brain barrier
BBTB	blood-brain tumor barrier
BCS	biopharmaceutics classification system
BDNF	brain-derived neurotrophic factor
BP	back-propagation
BUE	built-up edges
CA-4	combretastatin A-4
CL	clearance value
CM	carrier material
CMCs	ceramic matrix composites
CNN	convolutional neural network
CNS	central nervous system
CS	chain string
CSA	coconut shell ash
CSI	current source inverter
DDS	drug delivery systems
DI	direct injection
DILI	drug induced liver injury
DL	drug-likeness
DMSO	dimethyl sulfoxide
DOC	depth of cut
DR	diabetic retinopathy
DRC	diabetic retinopathy complications
DS	drug-score
DWT	discrete wavelet transformation
EATDA	2-ethylamino-1,3,4-thiadiazole
EDM	electro discharge machine
EGFR	endothelial growth factor receptor
EGT	exhaust gas temperature
ESIMS	electrospray ionization mass spectrometry

FDA	Food and Drug Administration
FR	feed rate
FSB	fixed carrier bandwidth
FTIR	Fourier transform infrared spectrometer
GIT	gastrointestinal tract
GLAST	glutamate-aspartate transporter
GLP-1	glucagon like peptide-1
HMEC-1	human microvascular endothelial
IA	interference alignment
IARC	International Agency for Research on Cancer
ICEF	iterative clipping and error filtering
ICI	inter-carrier interference
ICTF	iterative companding transform and filtering
IL	ignition lag
IM	intramuscular
IMP	inosine monophosphate
IMPDH	inhibitors of inosine monophosphate dehydrogenase
IR	intrarectal
ISI	inter-symbol interference
JB	jatropha biodiesel
KCS	kinematic chain string
KLS	kinematic link string
KLV	kinematic link value
LBM	laser beam machining
LDM	least distance matrix
LIMS	low-intensity Magnetic separation
LOM	light optical microscopy
MIMO	multiple-input multiple-output
MMCs	metal matrix composites
MODDS	modified oral release drug delivery system
MPAK	mitogen activated protein kinase
MQL	minimum quantity lubrication
MW	molecular weight
NGF	nerve growth factor
nHA	number of hydrogen bond acceptor
nHD	number of hydrogen bond doner
NLC	nanostructure lipid carrier
NPDR	non-proliferative diabetic retinopathy
NRTKs	non-receptor tyrosine kinases
NSCLC	non-small cell lung cancer
NT-4	neurotropin-4
NVD	neovascularization of the disk
OFDM	orthogonal frequency division multiplexing
PAPR	peak to average power ratio
PC	phosphatidylcholine
PC	prostate cancer
PCDDS	parenteral controlled drug delivery system

PD	Parkinson's disease
PD	partial discharges
PDR	proliferative diabetic retinopathy
PE	phosphatidylethanolamines
PID	proportional-integral-derivative
PKs	protein kinases
PMCs	polymer matrix composites
PPV	pars plana vitrectomy
PR	proliferative retinopathy
PS	phosphatidylserines
PVA	polyvinyl alcohol
PVP	polyvinylpyrrolidone
QoS	quality of service
RSM	response surface methodology
RTKs	receptor tyrosine kinases
SD	solid dispersion
SEM	scanning electron microscope
SLN	solid lipid nanoparticles
SM	solvent method
SPSS	statistical package for social sciences
TCVS	total chain value string
TDDS	transdermal drug delivery system
TKI	tyrosine kinase inhibitor
UTP	uridine triphosphate
VFI	voltage fed inverter
VSI	voltage source inverter
WEDM	wire electric discharge machining
WJM	water jet machining
WTDL	waste tire derived liquid
XRF	X-ray fluorescence

PREFACE

This book presents a selection of original and innovative works on computational optimization, modeling and simulation in the fields of renewable energy, sustainability, artificial intelligence (AI), operational research, management, mathematical modeling health care and pharmaceuticals, while also considering current environmental issues.

Modeling, simulation, and optimization are three critical components of modern design practice that have grown tremendously over the last few decades with the increased use of computer-aided designs in engineering and industry. This book takes a multidisciplinary approach to the trends and challenges of computational modeling, simulation, and optimization techniques and its applications in the field of engineering sciences.

The book focuses purely on computational modeling, simulation, and optimization techniques. It presents practical case studies and illustrations of computational techniques on materials and manufacturing in a diverse selection of applications, such as for drug delivery systems, for refrigeration models, for polymer and aluminum metal matrix composites, for automotive and aircraft applications, and even for talent management strategies on retaining talented employees.

Key features:

- Presents practical case studies and illustrations on materials and manufacturing, power and energy, heat transfer and thermal sciences, image processing, nanotechnology, biomedical engineering and wastewater technology
- Provides many applications of models, algorithms, and their fundamentals
- Discusses the theoretical foundations and analyses of computational modeling, simulation, and optimization techniques
- Includes detailed and comprehensive descriptions of applied techniques with mathematical modeling and application steps

Computational Optimization, Modeling, and Simulation for Engineering Applications will be of particular interest to researchers, engineers, industry professionals, and academicians and doctoral scholars in the major domains of engineering. Researchers and engineers as well as experienced experts will find it a handy reference.

The Organizing Committee of 7th International Conference Shaastrarth cordially expresses its thanks to all the authors and co-authors and all the diligent reviewers for their contributions to both the conference and this book.

—Editors

CHAPTER 1

DEVELOPMENT AND EVALUATION OF PIPERINE NANOSTRUCTURED LIPID CARRIER MORDANT DRUG DELIVERY SYSTEM

PRAVIN VERMA,[1] GITA SAHU,[1] and RITIKA SINGH[2]

[1]*Assistant Professor, Shri Rawatpura Sarkar University, Raipur, Chhattisgarh, India*

[2]*Assistant Professor, Rungta College of Pharmaceutical Science and Research, Raipur, Chhattisgarh, India*

ABSTRACT

In this research, we have developed and evaluated a piperine-containing mordant drug delivery system. We have formulated and characterized a PEGylated piperine nanostructured lipid carrier (NLC). The formulation was done by the fusion method. After that, we have also conducted a characterization process including SEM, TEM, FTIR, and *in vitro* as well as *in vivo* studies. The stability study of the optimized formulation, NLCO3, was carried out by evaluating physicochemical and *in vitro* parameters after storage for 1 month under freeze temperature, room temperature, and accelerated conditions are given in Table 1.2. The ointment was originated to be physically even at diverse temperatures, i.e., 8°C, 25°C and 45°C during and after storage period 1 month. There were no changes in the spreadability and no physical changes were observed. Moreover, no crystal growth; no phase separation was observed, indicating the stability of NLCO3. Although the average particle sizes were slightly increased as seen in SEM image of

Computational Optimization, Modeling, and Simulation for Engineering Applications.
Anupam Shukla, Sourabh Rungta, Mohan Awasthy, and Rakesh Himte (Eds.)

the formulation. The outcome of the stability study is existing in table from results it can be done that at different temperatures, i.e., 8°C, 25°C, and 45°C there was slightly but insignificantly decrease in percentage drug release. Results suggest that keeping the formulation in refrigerating condition minimizes stability problems of NLCs and ointment formulation.

1.1 INTRODUCTION

Vitiligo, also identified as leukoderma, is the most common chronic depigmentation or hypopigmentation disease, affecting approximately 1-4% of the world population. Around 50% of the cases are identified before the age of 20, and 25% before the age of 14. It is also considered as leukoderma. When injuries occur, such as cuts or burns, skin whitening can happen, but usually these patches can be converted into vitiligo. It causes damage to the melanocytes on the top surface of the epidermis and/or rarely the hair follicle, resulting in white patches on the skin. Leukoderma becomes more active over time. Generally, these diseases are not harmful to human beings medically, as they do not cause pain or cuts. However, their emotional and psychological effects can be distressing, especially if the visible areas of the body (such as the face, hands, arms, and feet) or the genitals are affected.

1.1.1 CATEGORIZATION OF VITILIGO

The classifications are based on partition, outline, and level of depigmentation. Segmental vitiligo (S) and non-segmental vitiligo (NS) are two different types of vitiligo. Based on other classification, three different types are found, i.e., localized, generalized, and universal vitiligo [1] (Figure 1.1).

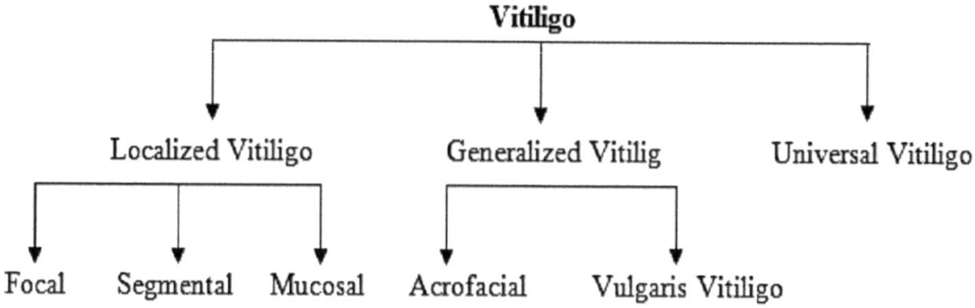

FIGURE 1.1 Classification of vitiligo.

1.1.2 NOVEL DRUG DELIVERY SYSTEM

The technique by which a drug is administered can have a vital result on its efficiency. A drug that shows minimum efficacy, but gives a higher therapeutic effect, and has low and high concentration benefits at all. From this, novel things on controlling the pharmacokinetics, pharmacodynamics, nonspecific toxicity, immunogenicity, bio-recognition, and efficiency of drugs were developed [2].

NDDS can add those according to physical characteristics and those depend on biochemical process. Physical mechanism also preferred as controlled drug delivery system includes osmosis, diffusion, erosion, dissolution, and electro transport [3]. Nanostructure lipid carrier (NLC) is second generation better drug carrier system having solid medium at room temperature. This carrier system is made up of biocompatible lipid, physiological, and biodegradable materials and surface-active agent and is usual by regulatory establishment for application in diverse drug delivery systems (DDS) [4] (Figure 1.2).

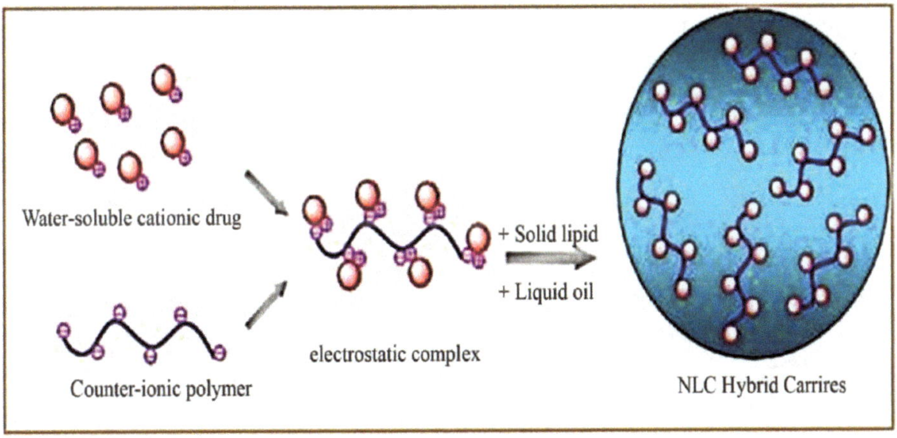

FIGURE 1.2 Nanostructure lipid carrier (NLC).

1.2 MATERIAL AND METHODS

1.2.1 MATERIALS

Piperine was purchased from Kshipra Pvt. Ltd. Indore (M.P.) India. The drug sample was identified and various preformulation parameters like solubility, partition coefficient, and quantitative estimation of piperine in a mixture of solvents were carried out. All these successful explorations paved the way for further studies.

1.2.2 FORMULATION AND CHARACTERIZATION

1.2.2.1 PREPARATION OF PEGYLATED PIPERINE

Polyethylene glycolated piperine was developed by combining emulsification and ultrasonication. Glyceryl monostearate and polyoxyethylene(40) stearate were liquefied at 70°C, combined together with soya-lecithin, soluble in DMSO (dimethyl sulfoxide) and it was cooled at 25°C. Piperine was mixed in DMSO and dissolved with the lipids to make up the organic phase. These phases were duly introduced into the aqueous phase, an aqueous solution of poloxamer 188, and were speedily stirred for around 10 minutes to achieve emulsification. The emulsion was subjected to ultrasonication using a probe sonicator (SONICS) and an ice bath and cooled temperature, to yield PEGylated piperine. The surfactant and DMSO were cleaned by diffusion across a dialysis membrane (MW cut-off 3.5 kDa). Piperine was freeze-dried and stored at 4°C [5]. According to requirement, piperine was redispersed [6, 7].

1.2.2.2 PREPARATION OF PEGYLATED PIPERINE LOADED NANOSTRUCTURED LIPID CARRIER (NLC)

The NLCs were developed by the melt dispersion ultrasonication method. The lipid tristearin and oleic acid, cholesterol, and phosphatidylcholine (soya lecithin) were mixed and heated at 75°C, along with the pegylated piperine, to form a uniform and clear oil phase. Meanwhile, the aqueous phase consisting of dispersing surfactant Tween-80 in double distilled water was maintained at 75°C [8]. Micro-emulsions were prepared with the help of the agitation method. Both the aqueous phase and oil phase were mixed at 600 rpm for 10 min. This hot micro-emulsion was diluted in cold water (2–3°C) below mechanical stirring to form NLC distribution such that the concentration of piperine in the final dispersion remains 2% w/w [9, 10] (Table 1.1).

TABLE 1.1 Formulation Table for PEGylated piperine Loaded NLCs

Sl. No.	Ingredients	Formulations Code					
		Nano-Lipid Carrier 1	Nano-Lipid Carrier 2	Nano-Lipid Carrier 3	Nano-Lipid Carrier 4	Nano-Lipid Carrier 5	Nano-Lipid Carrier 6
1.	Pegylated piperine (mg)	13	13	13	13	13	13
2.	Tristearin (mg)	200	200	200	150	150	150
3.	Oleic acid (mg)	50	75	100	50	75	100
4.	Cholesterol (mg)	0.015	0.015	0.015	0.015	0.015	0.015
5.	Soya lecithin (mg)	0.015	0.015	0.015	0.015	0.015	0.015
6.	Tween-80 (ml)	2	2	2	2	2	2

1.2.3 FORMULATION DEVELOPMENT

1.2.3.1 PEGYLATED PIPERINE NLCS LOADED OINTMENT FORMULATION BY FUSION METHOD

Three formulations nano-lipid carrier ointment NLCO1, nano-lipid carrier ointment NLCO2, and nano-lipid carrier ointment NLCO3 were performed as per the formula given in Table 1.2. Correct quantities of ingredients were weighed. The preparation was separated into two phases namely phase 1 (or oil phase) and phase 2 (or water phase). Ingredients of Phase I, i.e.,

TABLE 1.2 Composition of Various Ointment Formulations

Ingredients (% w/v)	NLCO1	NLCO2	NLCO3
Phase I or Oil Phase			
Cocoa butter	2	1	2
Coconut oil	5	5	2
PEG 6000	10	5	10
PEG 4000	20	15	20
Petroleum jelly	10	5	20
BHT	0.2	1	0.2
Tween 80	0.5	2	1
White bees wax	1	20	2
Phase II or Aqueous Phase			
Hard paraffin	1	10	5
Propyl glycol	1	1	1
Pegylated piperine loaded NLC	5	20	5
Rose water	50	2%	1%

coconut oil, polyethylene glycol 6000, polyethylene glycol 4000, petroleum jelly, were melted in a China dish using a hot plate. Dry powder (cocoa butter, BHT) was incorporated into the melted mixture followed by constant heating until absolute solubilization of powders in the oil phase successfully by adding together of Tween 80. Phase II (rose water, propylene glycol, and nanolipid structured carrier formulation) was melted in a separate 50 ml beaker to the smallest amount temperature so that all ingredients were separated accurately. Phase II was duly added into phase I, with continuous stirring until a smooth and uniform ointment was obtained.

1.3 RESULTS AND DISCUSSIONS

1.3.1 CHARACTERIZATION OF NLCS

A variety of process variables could change the preparation and properties of the NLC, and according to the preparation method, it was optimized and validated. The formulation of NLC containing piperine involves process variables. In the process of formulation development, the process variable is kept constant.

The mean particle size of NLC was analyzed by scanning electron microscopy and vesicle size established to be in the assortment of 60.14 to 70.55 (Table 1.3). For Entrapment efficiency was extra when span 80 was utilized as a surfactant (70.55±1.6, NLCs 5) as compared to other preparations. From the images of scanning electron microscopy and transmission electron microscopy, the sizes of the vesicles were approximately the same before (66.97 μm) and after extrusion (67 μm) through a membrane of pore size (12 μm) lesser than that of the vesicles. Thus, we declare that the preparation can significantly easily infuse through skin pores (20 μm). Span 80 analyzed extra entrapment efficiency than Tween 80 (Table 1.3). This is because span 80 is less hydrophilic (lower HLB) than Tween 80, as piperine is hydrophobic, span 80 shows extra entrapment efficiency. *In vitro* release studies were recorded for a 2-hour period. NLC preparations give drug release 66 (NLC 1), 62 (NLC 2), 80 (NLC 3), 68 (NLC 4), 84.17 (NLC 5), 81.9 (NLC 6), respectively. FTIR spectra of pegylated piperine loaded NLC (Figure 1.4) showed stretching of functional groups N-H (3429.7 cm^{-1}), aromatic C-H (3088.2 cm^{-1}), alkenes C-H (3008.2 cm^{-1}) and aliphatic C-H (2937.1 cm^{-1}) groups, respectively. That there were no interactions occurred between the piperine and other excipients.

1.3.2 VESICLE MORPHOLOGY

Morphology and size distribution of NLC were obtained by electron microscopy and SEM (scanning electron microscopy). To establish the flexible penetration nature of NLC, the vesicle suspension was prepared to pass through membrane filters of 12 μm pore size which were situated above a Buchner funnel and saturated membrane was carefully scraped and scanning electron microscopy was taken (Figures 1.3 and 1.4).

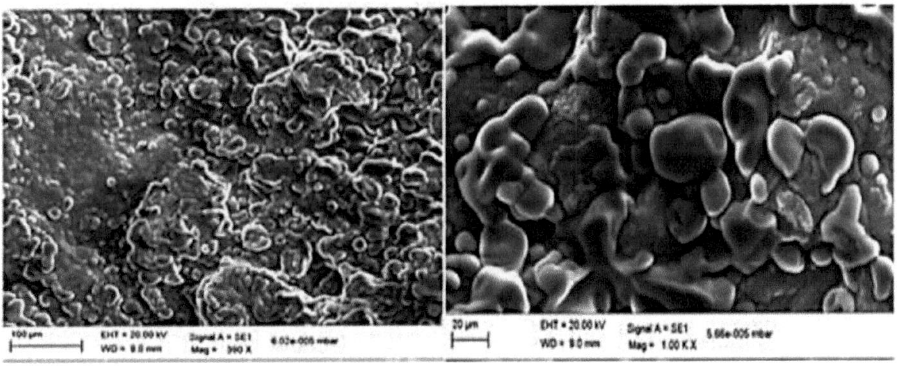

FIGURE 1.3 SEM image of ointment.

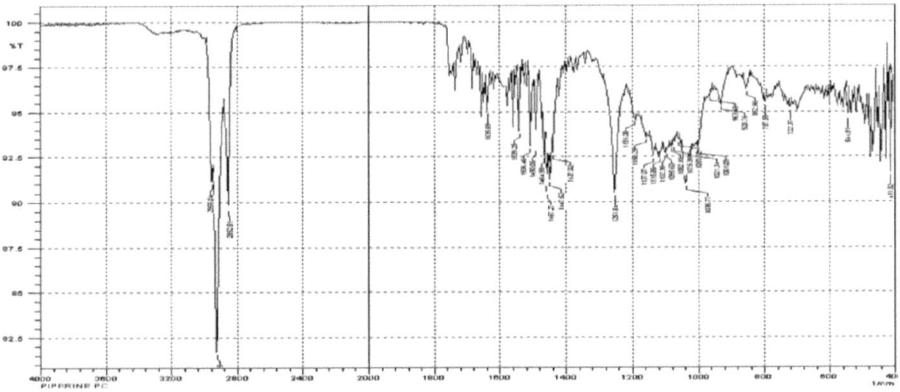

FIGURE 1.4 FT-IR study of pegylated piperine loaded NLCs complex.

1.3.3 ENTRAPMENT EFFICIENCY

The piperine incorporated NLCs were centrifuged at 14,000 rpm for 30 minutes. The supernatant (1 ml) was taken and diluted with phosphate buffer (pH 6.6). The quantity of unentrapped piperine was analyzed by UV spectrophotometer at 340 nm with appropriate dilutions. Encapsulation efficiency is expressed as the percentage of drug entrapped.

$$\text{Percentage Entrapment efficiency} = \frac{\text{Total amount of drug} - \text{un entrapped drug}}{\text{Total amount of drug}} \times 100$$

TABLE 1.3 Characterization of Pegylated Piperine Loaded Nanostructure Lipid Carrier

Batch Code	Mean Vesicular Size (μm)	Entrapment Efficiency	In Vitro Drug Release
NLCs 1	64.01±1.36	63.39±0.01	66%
NLCs 2	64.56±1.90	53.01±0.11	62%
NLCs 3	66.96±2.19	69.05±1.04	80%
NLCs 4	60.14±0.12	60.19±0.02	68%
NLCs 5	70.55±1.62	80.23±1.04	84.17%
NLCs 6	67.11±1.91	71.20±1.02	82%

Microscopic characterization of pegylated piperine loaded NLCs by fluorescence and transmission electron microscopy.

The surface morphology, i.e., shapes and size of the blank and pegylated piperine loaded NLCs, was observed by fluorescence microscopy and transmission electron microscopy. NLC preparations were diluted with distilled water. One drop of the diluted suspension was then directly placed on the film grid, stained by 1% aqueous preparation of phosphotungstic acid, and examined after drying. A thin film of NLCs was spread on a slide and a coverslip was placed over it, then observed under the electron microscope. The images of both are presented in Figure 1.5.

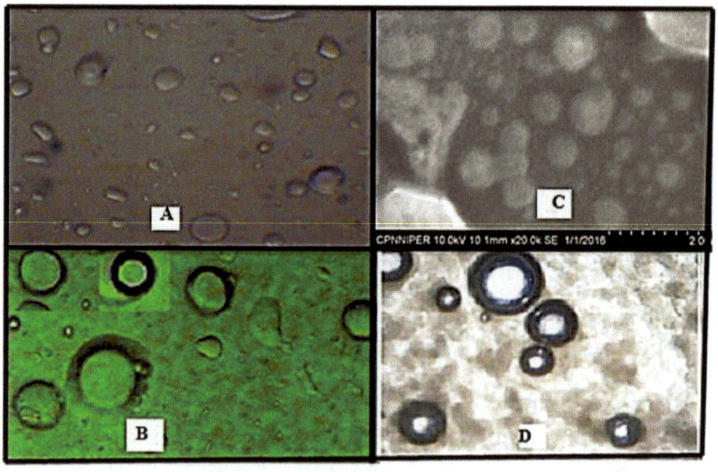

FIGURE 1.5 Fluorescence images of (A) blank NLCs; (B) pegylated piperine loaded NLCs; TEM images of (C) blank NLCs; and (D) pegylated piperine loaded NLCs.

1.3.4 IN VITRO DRUG RELEASE

The *in vitro* release of pegylated piperine loaded NLCs was done by taking NLCs containing an equivalent of 5 mg of the drug, which was placed in the opening end of a test tube as a donor compartment. The donor compartment was covered with a cellophane membrane (molecular weight cut off 12,000 Da). The receptor solution was 100 ml of phosphate buffer (pH 6.6), which was maintained at 37±0.2°C and continually stirred at 100 rpm in a thermostatically constant water bath shaker. About 2 ml of sample was withdrawn from the solution compartment at 15-minute intervals and immediately made up with an equal volume of fresh medium. Experiments were performed three times for each preparation, and sink conditions were always maintained throughout the whole procedure. All formulations were analyzed for piperine load spectrophotometrically at a wavelength of 340 nm (Figure 1.6).

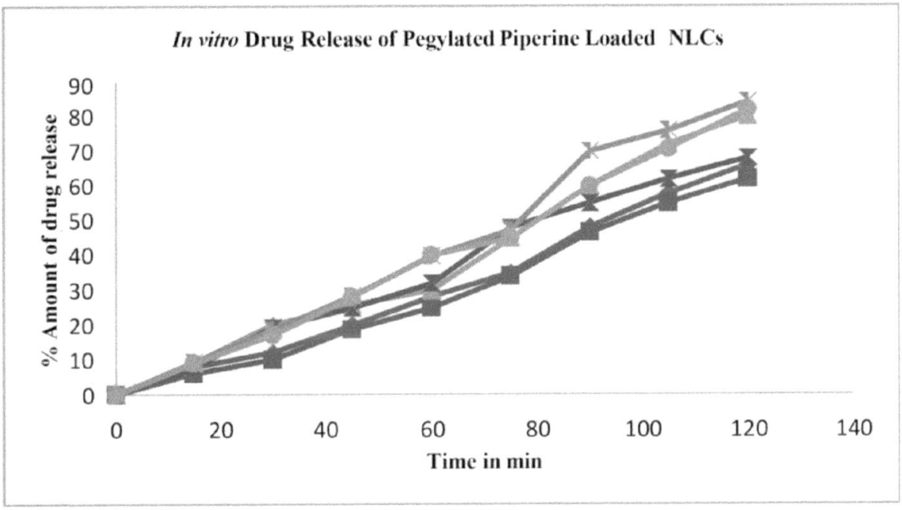

FIGURE 1.6 *In vitro* drug release of pegylated piperine loaded NLCs.

1.3.5 IN VITRO DRUG RELEASE AND RELEASE KINETICS OF SELECTED FORMULATION

The result of dissolution profiles showed that regarding the formulation, the values of drug release after 1.5 hr. were established to be 84.17% piperine from NLCs 5, and 85% piperine from optimized ointment (NLCO3) (Figure 1.7).

On comparing the release of piperine from optimized ointment formulation (i.e., NLCO3) with pegylated piperine loaded NLC formulation (i.e., NLC 5), it was clearly observed that the percentage drug release was slowed down until about 90 minutes.

The released study data were treated for zero order, first order, and Higuchi model patterns for determination of drug release during the dissolution process. The best fitting kinetic model with the higher correlation was found with the Higuchi model. The regression coefficients of the optimized formulation were 0.950 for the zero order equation, 0.959 for the first order equation, and 0.981 for the Higuchi model, respectively.

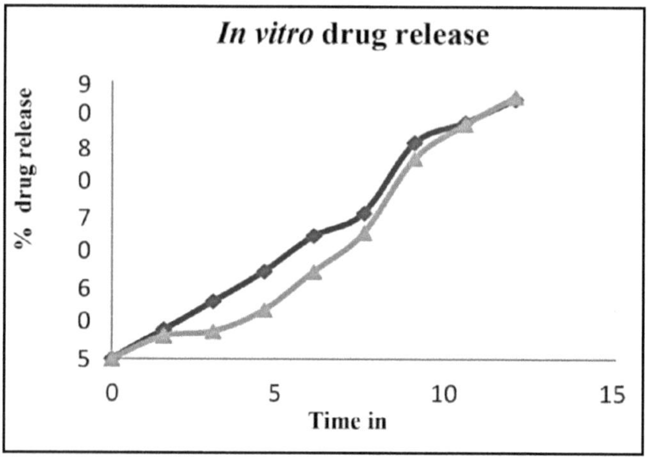

FIGURE 1.7 *In vitro* drug release.

1.4 RESULT AND DISCUSSION

The stability study of optimized formulation that is $NLCO_3$ was carried out by evaluation of physicochemical and *in-vitro* parameters after storage for 1 month under freeze temperature, room temperature and accelerated conditions. The ointment was found to be physically stable at different temperatures, i.e., 8°C, 25°C, and 45°C during and after storage period 1 month. There were no changes in the spreadability, and no physical changes were observed. Moreover, no crystal growth; no phase separation was observed indicating the stability of $NLCO_3$. Although the average particle sizes were slightly increased as seen in SEM image of the formulation. The results of the stability study are represented as it can be accomplished that at various temperatures, i.e., 8°C, 25°C, and 45°C there was slightly but insignificantly decrease in percentage drug release. Results suggest that keeping the formulation in refrigerating condition minimizes stability problems of Nanostructured Lipid Carrier and ointment formulation

KEYWORDS

- ***in vitro* parameters**
- **nanostructured lipid carrier**
- **novel drug delivery system**
- **optimization**
- **physiochemical particle size**
- **spreadability**
- **stability**

REFERENCES

1. Gupta, B., Ramasamy, T., Poudel, B. K., Choi, H. G., Yong, C. S., & Kim, J. O., (2017). Development of bioactive PEGylated nanostructure platforms for sequential delivery of doxorubicin and imatinib to overcome drug resistance in metastatic tumors. *ACS Applied Materials & Interfaces*, 1–39.
2. Iqbal, M. A., Md, S., Sahni, J. K., Baboota, S., Dang, S., & Ali, J., (2012). Nanostructure lipid carriers system: Recent advances in drug delivery. *J. Drug Target, 20*(10), 813–830.
3. Jain, P., Mishra, A., Yadav, S. K., Patil, U. K., & Baghel, U. S., (2009). Formulation development and characterization of solid lipid nanoparticles containing nimesulide. *International Journal of Drug Delivery Technology, 1*, 24–27.
4. Kharat, A., & Pawar, P., (2014). Novel drug delivery system in herbals. *Int. J. Pharm. Chemical and Biol. Sci., 4*, 910–923.
5. Manivannam, R., & Kugalur, G. P., (2010). Recent advances in novel drug delivery system, International *Journal of Research in Ayurveda & Pharmacy, 1*(2), 316–326.
6. Minder, E. I., (2010). Afamelanotide, an agonistic analog of α-melanocyte stimulating hormone, in dermal photo toxicity of erythropoietic protoporphyria. *Expert Opin. Investig. Drugs, 19*, 1591–1602.
7. Mishra, P., Nayak, B., & Dey, R. K., (2016). PEGylation in anti-cancer therapy: An overview. *AsianJournal of Pharmaceutical Sciences, 11*, 337–348.
8. Noha, N. D., Arwa, M. H., & Shereen, F. G., (2011). Topical antioxidant and narrowband versus topical combination of calcipotriol plus betamethathone dipropionate and narrowband in the treatment of vitiligo. *J. Life Sci., 8*, 186–197.

9. Ongenae, K., Van, G. N., & Naeyaert, J. M., (2003). Evidence for an autoimmune pathogenesis of vitiligo. *Pigment Cell Res., 16*(2), 90–100.
10. Pei, T., Zheng, C., Huang, C., Chen, X., Guo, Z., Fu, Y., Liu, J., & Wang, Y., (2016). Systematic understanding the mechanisms of vitiligo pathogenesis and its treatment by Qubaibabuqi formula. *Journal of Ethanopharmacology*, JEP10203.

CHAPTER 2

A REVIEW ARTICLE ON BY-PRODUCT OF BLAST FURNACE PROCESS: FLUE DUST

SATYENDRA KUMAR DEWANGAN[1] and AJAY VIKRAM AHIRWAR[2]

[1]*Research Scholar, Department of Civil Engineering, National Institute of Technology, Raipur, Chhattisgarh, India*

[2]*Assistant Professor, Department of Civil Engineering, National Institute of Technology, Raipur, Chhattisgarh, India*

ABSTRACT

The steel industry plays a crucial part in the progress and development of India by contributing approximately 2% to the National GDP. Every steel industry operates on the Blast Furnace principle, which chemically reduces the raw material and substantially changes iron oxides into liquid iron, also acknowledged as "hot metal." In this review, a major by-product of blast furnace called "Flue dust" is reviewed for its generation, properties, characterization, component, and current applications for its utilization or management. The different environmental impacts caused by Flue Dust are also highlighted in this study.

2.1 INTRODUCTION

At present, India with its 3.250 trillion nominal economies (2022) is the 6 th largest economy in the world which has its footprint in almost every sector. One of the irreplaceable sectors is the Steel sector which contributes almost 2% of the GDP. The Indian steel industry has surpassed Japan in

Computational Optimization, Modeling, and Simulation for Engineering Applications.
Anupam Shukla, Sourabh Rungta, Mohan Awasthy, and Rakesh Himte (Eds.)

the year 2019 and became the world's 2 nd largest steel producer. A blast furnace is the heart of every steel industry as it performs the major role in chemical reduction by converting the iron ore into liquid pig iron called "Hot metals." This blast furnace operation is also responsible for the generation of by-products like water-cooled slag, air cool slag, sludge, and flue dust, etc. This chapter majorly emphasizes on the review of the flue dust in order to explore the process of its generation, elements involved in it, different ways of its utilization its accumulation and different ways of its application in the various sectors. The chapter reviews the Flue dust on the basis of its properties, characterization, components, and existing utilization techniques. The review of these parameters from different secondary sources will facilitate the future study of effective utilization of Flue dust specifically in the construction field (Figure 2.1).

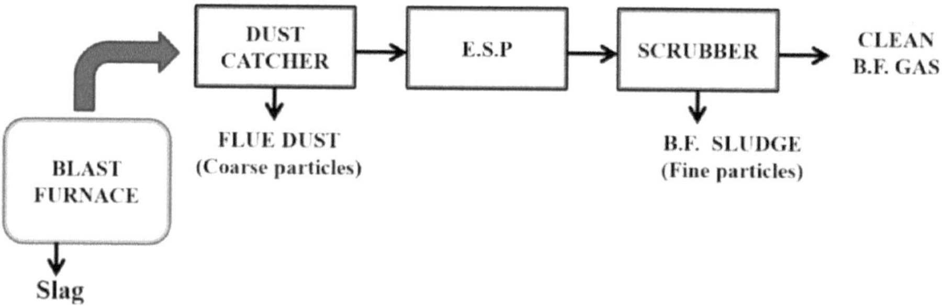

FIGURE 2.1 Cleaning of blast furnace gas.

2.2 GENERATION OF FLUE DUST

In the blast furnace operation, the major raw material like iron ore, sinter along with flux like limestone, dolomite, and different fuels like coke is fed from the top opening of the blast furnace through charging devices. For the combustion of fuels, hot air which consists of oxygen is blown from the tuyere (pipe-like structure) of the furnace which converts the coke into carbon monoxide and also reduces the iron oxide of the main raw material, i.e., iron ore [1]. As a result of this process, hot metal is produced in the liquid form but with some impurities or byproducts. These by-products consist of elements such as silicon, carbon, sulfur, phosphorus, etc. Slag and other gangue materials are also produced as impurities that are separated from the base opening of the blast furnace.

The flue gases exit from the upper passage of the furnace. This flue dust generally contains lighter material like coke, fluxes, coal, iron ore, which passes from the dust catcher device containing cyclone at the bottom [5]. This device separates the high-density material from the flue gases which is called flue dust. The separated gas is then passed through primary and secondary washers which clean the gases. This blast furnace gas has a high calorific value and is highly combustible which is used as a fuel for another integrated system [1].

2.3 CHARACTERIZATION OF FLUE DUST

The blast furnace operation is very complex and hence a small change in the proportion of materials used may lead to an increase in the production of impurities. In order to continue balanced and effective operation, it is essential to determine and observe the behavior of fine particles generated from the flue gases (flue dust), which can be achieved through the characterization test of the flue dust sample.

2.3.1 *METHODS USED*

- XRD (X-ray powder diffraction);
- SEM (scanning electron microscopy);
- LOM (light optical microscopy);
- XRF (X-ray fluorescence);
- FTIR (Fourier transform infrared spectrometer).

2.3.2 *SAMPLE PREPARATION*

The sample is prepared by dry sieving using 1, 0.5, 0.25, 0.125, 0.075, 0.063, and 0.045 mm of square sieves. And Chemical analysis has been done by SSAB.

2.3.3 *INVESTIGATION AND RESULT OF FLUE DUST*

The above-mentioned methods were employed to analyze the sieved sample of flue dust. The result shows that:

- The size fraction of <0.5 mm has the predominant amount of flue dust [2].
- A higher ore-to-coke ratio shows immense requirement on burden strength for all charged raw materials that descend through the blast furnace shaft [23].

- Principal components observed in flue dust are NaCl, KCl, Fe_2O_3 and carbon [4].

The main elements in the flue dust which contains metal are (Table 2.1):

- Carbonaceous materials, i.e., the fine char and unconsumed coke carried by the dust [4].
- Particles that contain carbon dominated in the fractions greater than 0.075 mm are found to have mainly coke particles [2].
- Fe_2O_3 (Fe-containing particles) has formed the main component of flue dust fractions < 0.063 mm [2].
- Ratios of mass of chars and mass of cokes to the carbonaceous element in flue dust are calculated as 90.56% and 9.44% respectively [4].
- The characterization also helps to determine the actual source of carbon in flue dust which is from the carbonaceous matter charged to the blast furnace. It can also be inferred that the carbon formation is not due to the fluxes fed up with charging [3].

TABLE 2.1 Characterization of Flue Dust

Elements	Mass Fraction (%)
Fe	33.58
K	13.8
Na	8.46
Zn	7.04

Source: Adapted from: Energy and Fuels, *31*(8), 8415–8422.

2.3.4 *IMPORTANT POINTS ABOUT FLUE DUST*

- Total Fe percentage in flue dust varies from 20-40 wt.% and C content is determined between 30 and 50 wt.% [24-26].
- Elements such as Si, Ca, Mg are present in insignificant amounts.
- From the findings of research conducted in an Egyptian iron and steel industry for characterization of flue dust, it has been observed that around 74 wt.% of the component was under a grain size of 210 mm [27].
- Chemical investigation of flue dust demonstrated that carbon is present in the coarser sizes which are over 210 mm and iron is

present in the finer size under 210 mm. This determination showed that almost each and every material in the flue dust was in the size of < 210 mm [28].

2.4 PROPERTIES OF FLUE DUST

The United States Steel Corporation (an integrated steel producer having the majority of its operations in central Europe and North America) has described the properties of blast furnace flue dust in their 2017 safety data sheet, which is listed in Tables 2.2 and 2.3.

TABLE 2.2 Size Distribution of Flue Dust

Sample/Origin	Grain Size Distribution (mm, %)						
	<0.08	0.08–0.2	0.2–0.8	0.8–1.25	1.25–2	2–5	>5
Dust bag	8.60	10.43	15.64	7.67	11.65	23.01	23.00
Dust cyclone	17.32	46.73	34.64	0.98	0.33	–	–
Sludge bin	7.78	10.20	20.28	15.56	32.53	13.35	–

Source: Adapted/reconstructed from: Ref. [12] ©2014, U.P.B. Sci. Bull., Series B, *76*(1).

TABLE 2.3 Properties of Flue Dust (Original)

Properties	Description
Hazardous	• Carcinogenic and toxic (may cause cancer and damage to lungs and skin from repeated exposure);
	• Harmful if swallowed;
	• Cause irritation in eyes and skin;
	• Hazardous for the environment if dumped or burned.
Composition	• Carbon;
	• Quartz (crystalline silica);
	• Iron and iron oxides;
	• Metallic silicates.
Appearance	• Fine black powder;
Flammability	• Nonflammable;
Solubility	• Negligible;
Size	• The size distribution shows that around 20% are from sludge and dust bag.
	• Around 65% of the cyclone flue dust particles have a median diameter less than 200 μm [7].

2.5 RECYCLING OR UTILIZATION OF FLUE DUST

The byproduct of blast furnaces, such as sludge and flue dust, is mostly recovered from the steel industry by recycling in industrialized countries through sinter making. This recycling process is generally managed or controlled, counting on the study of the waste material [29]. The availability of different methods developed and introduced over time to reduce pollution and to reuse the byproducts demonstrates once again the need for further determination of different kinds of waste using a more extensive and detailed way of in-depth investigation and characterization.

The research conducted over the last three decades has elaborated on the different methods of reuse and utilization of by-products, powdered wastes. Some of these methods are listed below:

- Sintering [30];
- Reduction of size [31];
- Sludge leachability [32–34];
- Pelleting by sintering of the furnace dust [35];
- Reduction of iron oxides [36];
- Reduction melting in a rotary kiln [37].

Direct recovery of iron by a fluidized bed [38] involves the simultaneous injection of coal and blast furnace dust [39]. The pollution that consists of metallurgical industrial wastes has been extensively investigated by observing and studying their environmental risks and by developing an in-depth scientific interest in utilizing, recovering, and recycling the byproducts after conducting chemical and micro-morphological characterization [40], which was developed either in Romania [41–43]. Generally, the flue gases that contain flue dust involve the process of evicting bigger particulates by the use of a device called a dry dust catcher, followed by wet cleaning of the gas for the removal of fine particles.

These residues primarily consist of iron, calcium, silicon, aluminum, magnesium, and also carbon in the form of coke. This waste contains a significant number of heavy metals which are harmful to the environment if dumped or used for landfills. Alongside these heavy metals, they can also be separated and utilized in order to make them useful for other applications.

The obtained dust and sludge mostly contribute 4 to 7% of the hot metal produced having iron percentage of 15-40%, the carbon content of 25-40%, and 7-10% SiO2. This dust which contains a good proportion of heavy metals can be used properly to increase the manufacturing capacity of the blast furnace system and to prevent environmental damages or pollution. Some proportions of coarse Flue dust and sludge among these wastes are utilized in sintering and pelletizing blends.

2.5.1 RECYCLING OF ZINC FROM FLUE DUST

The metal element Zinc contributes approximately 7% of flue dust, which can be recycled and utilized by a zinc production mechanism that uses NH_3-CH_3COONH_4-H_2O as a leaching agent. The reaction of CH_3COONH_4 and NH_3–H_2O system promotes and boosts zinc extraction, which dissolves approximately 77.79% of Zn. As a result of electrospray ionization mass spectrometry (ESIMS) analysis and Fourier transforming infrared spectrum (FT-IR), zinc can be added with Carboxylate anion to form Zn-complexes [8]. According to the research, it has been observed that the Zn Content in Flue dust increases with a decrease in the particle size [9]. This Zn can be separated through the Air classification technique in Zn enriched flue dust, which separates a small fraction of dust (10–20%) and leads to the reduction of 40% to 60% of Zn with a minimal carbon loss of 5–10% [9].

2.5.2 RECOVERY OF CARBON FROM FLUE DUST

In order to recover the carbon content in flue dust, the flotation technique can be employed, which shows that around 90% of the carbon can be recovered with a carbon grade between 70% and 80% by using diesel oil as a collector MIBC (methyl isobutyl carbinol) as a FROTHER [11] (Table 2.4).

TABLE 2.4 Recovery of Carbon Values by Column Flotation Technique

Sl. No.	Air Rate (liter/min)	Rourkela Steel Plant		Tata Steel Plant	
		Carbon (%)	Carbon Rec. (%)	Carbon (%)	Carbon Rec. (%)
1.	3.5	85.2	34.3	–	–
2.	4.5	84.9	60.3	–	–
3.	5.0	82.9	72.3	80.0	97.2
4.	6.0	81.3	84.0	80.2	98.9
5.	6.5	79.0	95.3	80.7	99.0
6.	7.0	76.4	94.2	79.6	91.2
7.	7.5	75.0	95.0	78.2	89.2

Source: Adapted/reconstructed from: Ref. [6]. The European Journal of Mineral Processing and Environmental Protection, 2(2), 61–68.

2.5.3 RECOVERY OF IRON FROM FLUE DUST

The iron-rich elements/grains such as hematite, wustite, magnetite, and Fe metals can be separated from the Flue Dust by a method called LIMS (low-intensity magnetic separation) [14]. It can be inferred from the paper that approx. 50-56% of overall iron recovery can be possible from this technique [11].

2.5.4 RECOVERY OF ALKALI CONTENT FROM FLUE DUST

The alkali content in the flue dust can be removed by carrying out the techniques such as scrubbing and washing. It was found that 85% of Na and 60% of K, which are harmful alkalis, can be removed by this process (Table 2.5).

TABLE 2.5 Elimination of Alkalis from Flue Dust by Scrubbing

Solid Liquid Ratio	Scrubbing Time (Min)	Particle Size of Sample (μm)	Na Elimination (%)	K Elimination (%)
1:1	30	As received	75.1	21.6
1:1.5	30	As received	72.0	34.9
1:2	30	As received	70.9	29.3
1:1	60	As received	76.3	22.0
1:1	240	As received	80.3	32.1
1:1	30	–300	84.4	60.0
1:1.5	30	–300	85.5	57.5
1:1	30	–150	95.0	60.2
1:1.5	30	–150	95.1	60.1

Source: Adopted/reconstructed from: Ref. [6]. The European Journal of Mineral Processing and Environmental Protection, 2(2), 61–68.

2.5.5 REDUCTION OF MANGANESE ORE FROM FLUE DUST

The coke with a carbon content of 83.49% along with manganese ore concentrate and scale dust from the mill are heated at 1,200°C in a tube type furnace located in argon atmospheres for 2 hrs. This gives a reduction of 67.9% by using 200% of theoretical carbon content [12].

After this the same experiment is carried out by substituting the mill scale dust with flue dust having 23% carbon and 34.64% hematite, as a carbon source and iron. This modification gives an encouraging result with an increase in the reduction degree by 12%, i.e., 68.35% [12].

2.5.6 FLUE DUST RECOVERY

Flue dust is hazardous for the environment if dumped or burned and hence can be recovered by the application of hydrometallurgical techniques and solvent extraction. Flue dust can be leached using a variety of leaching reagents, including complexing, acidic, basic, and pregnant solutions [10].

2.5.7 FLUE DUST AS A ROAD ENGINEERING

This high asphalt content can be attributed to the flue dust's high absorption capacity and texture. In general, design-wise, blend of flue dust and slag percentages are technically attainable. Even if these are vulnerable to moisture and saturation, these limitations can be overcome by using specialized additives to improve the adhesiveness of asphalt mixtures. Because the byproducts slag and flue dust comply with all of the demands for asphalt mixtures, their use in the production of asphalt mixtures for road construction is feasible. This will also reduce pollution and environmental damage caused by open-air dumping.

2.6 CONCLUSION

The steel industry works on the principle of blast furnace operation, which takes main raw materials such as iron ore, sinter, and flux like limestone, dolomite, and fuels like coke as input from its charging end and converts it into the liquid iron called "hot metals." This process is also responsible for the production of various by-products, one of which is flue dust. This chapter examines and analyzes each and every parameter of flue dust, such as the generation of flue dust, flue dust components, its characterization, properties, recovery methods, utilization techniques, etc. The review of papers facilitates the future research of effective utilization of flue dust in the construction sector and hence helps to develop new efficient techniques of flue dust utilization.

KEYWORDS

- **alkali**
- **blast furnace**
- **by-product**
- **flue dust**
- **hot metal**
- **raw material**
- **road engineering**

REFERENCES

1. Ashrit, S. S., Chatti, R. V., & Sarkar, S., (2019). Identification of the carbon source in blast furnace flue dust through characterization and statistical analysis. *International Journal of Environmental Analytical Chemistry, 101*(10), 1–15. https://doi.org/10.1080/03067319.2019.1682142.
2. Basu, G. S., Sharma, R. P., & Dhillon, A. S., (1996). *Dust Collecting Technologies for Blast Furnace and Steel Melting Shops,* 129–139.
3. Caravaca, C., Cobo, A., & Alguacil, F. J., (1994). Considerations about the recycling of EAF flue dusts as source for the recovery of valuable metals by hydrometallurgical processes. *Resources, Conservation and Recycling, 10*(1, 2), 35–41. https://doi.org/10.1016/0921-3449(94)90036-1.
4. Coedo, A. G., Dorado, T., Padilla, I., Maibusch, R., & Kuss, H. M., (2000). Slurry sampling electrothermal vaporization inductively coupled plasma mass spectrometry for steelmaking flue dust analysis. *Spectrochimica Acta, Part B: Atomic Spectroscopy, 55*(2), 185–196. https://doi.org/10.1016/S0584-8547(99)00179-2.
5. Cramp, G. B., (1921). Blast furnace flue dust recovery. *Scientific American, 3*(6supp), 543–546. https://doi.org/10.1038/scientificamerican06011921-543supp.
6. Das, B., Prakash, S., Reddy, P. S. R., Biswal, S. K., Mohapatra, B. K., & Misra, V. N., (2002). Effective utilization of blast furnace flue dust of integrated steel plants. *The European Journal of Mineral Processing and Environmental Protection, 2*(2), 61–68. Retrieved from: http://www.ejmpep.com/das_et.al.pdf (accessed on 21 December 2023).
7. Furnace, B., & Dust, F., (2020). *Section 1 – Identification; Section 2 – Hazard (s) Identification; Section 3 – Composition/Information on Ingredients; Section 4 – First-aid Measures; Section 5 – Fire-fighting Measures; Section 6 – Accidental Release Measures, 14576*(c), 1–4.
8. Jascisak, J., Havlik, T., Hoang, T. Z., & Turek, P., (2015). Hydrometallurgical processing of blast furnace flue dust containing zinc. *METAL 2015 – 24th International Conference on Metallurgy and Materials, Conference Proceedings* (pp. 1289–1293).
9. Lanzerstorfer, C., (2016). Air classification of blast furnace dust catcher dust for zinc load reduction at the sinter plant. *International Journal of Environmental Science and Technology, 13*(2), 755–760. https://doi.org/10.1007/s13762-015-0903-1.
10. Lundgren, M., Leimalm, U., Hyllander, G., Ökvist, L. S., & Björkman, B., (2010). Off-gas dust in an experimental blast furnace part 2: Relation to furnace conditions. *ISIJ International, 50*(11), 1570–1580. https://doi.org/10.2355/isijinternational.50.1570.
11. Ma, A., Zheng, X., Zhang, L., Peng, J., Li, Z., Li, S., & Li, S., (2018). Clean recycling of zinc from blast furnace dust with ammonium acetate as complexing agents. *Separation Science and Technology (Philadelphia), 53*(9), 1327–1341. https://doi.org/10.1080/01496395.2018.1444057.
12. Mihaiescu, D. C., Predeanu, G., & Panaitescu, C., (2014). Characterization of some blast furnace waste dusts. *UPB Scientific Bulletin, Series B: Chemistry and Materials Science, 76*(1), 227–234.
13. Morgan, G., & Davies, G. R., (2010). Germanium and gallium in coal ash and flue dust. *Journal of the Society of Chemical Industry, 56*(32), 717–721. https://doi.org/10.1002/jctb.5000563202.

14. Özdemir, İ., Şeşen, F. E., & Şeşen, M. K., (2016). *Usability of Blast Furnace Flue Dust in Solid-State Reduction of Manganese Ore Concentrate in the AIM of Ferromanganese Production*, 994–997.
15. Pitt, G. J., & Fletcher, M. F., (Late) (1955). The spectrographic determination of germanium in coal and flue dust. *Spectrochimica Acta*, *7*(C), 214–218. https://doi.org/10.1016/0371-1951(55)80066-5.
16. Rath, S. S., Rao, D. S., Tripathy, S. K., & Biswal, S. K., (2018). Characterization vis-á-vis utilization of blast furnace flue dust in the roast reduction of banded iron ore. *Process Safety and Environmental Protection*, *117*, 232–244. https://doi.org/10.1016/j.psep.2018.05.007.
17. Robinson, R., (2005). High temperature properties of by-product cold bonded pellets containing blast furnace flue dust. *Thermochimica Acta*, *432*(1), 112–123. https://doi.org/10.1016/j.tca.2005.04.015.
18. Uys, J. M., & Kirkpatrick, J. W., (1963). The beneficiation of raw materials in the steel industry and its effect upon air pollution control. *Journal of the Air Pollution Control Association*, *13*(1), 20–32. https://doi.org/10.1080/00022470.1963.10468136.
19. Xing, X., Chen, Y., & Liu, Y., (2018). Study of the reduction mechanism of ironsands with addition of blast furnace bag dust. *Metallurgical Research and Technology*, *115*(2). https://doi.org/10.1051/metal/2017097.
20. Zhong, Y., Qiu, X., Gao, J., & Guo, Z., (2017). Structural characterization of carbon in blast furnace flue dust and its reactivity in combustion. *Energy and Fuels*, *31*(8), 8415–8422. https://doi.org/10.1021/acs.energyfuels.7b01146.
21. Dancy, T. E., Sadler, A. T., & Lander, H. N., (1958). Process analysis of blast furnace operation with oxygen and steam. *Blast Furnace Coke Oven and Raw Materials Proceedings* (p. 16).
22. Vreeland, G. W., (1916). *The Distribution of Raw Materials in the Blast Furnace* (p. 106). /1/S/ Yearbook.
23. Loo, C. E., & Bristow, N., (1998). *Ironmaking Steelmaking, 25*, 222.
24. Das, B., Prakash, S., Reddy, P. S. R., & Misra, V. N., (2007). *Res. Conserv. Recycl., 50*, 40.
25. Robinson, R., (2005). *Thermochim. Acta, 432*, 112.
26. Sikström, P., & Ökvist, L. S., (2002). *TMS fall 2002 Extraction and Processing Division Meeting on Recycling and Waste Treatment in Mineral and Metal Processing, TMS, Warrendale, PA and Luleå University of Technology, Luleå and MEFOS Met. Res. Inst., Luleå*, 119.
27. Yehia, A., & El-Raheim, F. H., (2005). *Miner. Process. Extra. Metall., 114*, 207.
28. Asadi, Z. B., Mowla, D., Shariat, M. H., & Fathi, K. J., (1997). *Hydrometallurgy, 47*, 113.
29. Das, P. S., Reddy, P. S. R., & Misra, V. N., (2007). An overview of utilization of slag and sludge from steel industries. *Resources, Conservation and Recycling*, *50*, 40–57.
30. Ellis, J. D., (1999). Agglomeration or injection—Two cycling alternatives. In: *Learn Strategies for Coping with Steel Mill Wastes and Profiting from by-Products* (Vol. 5).
31. Das, P. S., Reddy, P. S. R., Biswal, S. K., Mohapatra, B. K., & Misra, V. N., (2002). Effective utilization of blast furnace flue dust of integrated steel plants. *European Journal of Mineral Processes and Environmental Protection*, *2*(2), 61–67.
32. Piecuch, T., & Dabrowska, L., (1983). Techno-economic prospects of utilizing wet-cleaned blast-furnace dust. *Hutnik (Katowice), 50*(5), 174–176.

33. Zeydabadi, B. A., Mowla, D., Shariat, M. H., & Fathi, K. J., (1997). Zinc recovery from blast furnace flue dust. *Hydrometallurgy, 47*(1), 113–125.
34. Van, H. P., Vandecasteele, C., Swennen, R., & Mortier, R., (2000). Zinc and lead removal from blast furnace sludge with a hydrometallurgical process. *Environmental Science and Technology, 34*(17), 3802–3808.
35. Uno, S., Umetsu, Y., Ohmizu, M., & Munakata, S., (1979). Dezincing equipment and operation based on wet classification of wet-cleaned BF dust. *Nippon Steel Tech Rep, 13*, 80–85.
36. Fosnacht, D. R., (1985). Characterization and utilization of iron-bearing steel plant waste materials. In: *Physical Chemistry of Extractive Metallurgy Conference* (pp. 479–492).
37. McClelland, J. M., & Metius, G. E., (2003). Recycling ferrous and nonferrous waste streams with FASTMET. *JOM J. Min., Met., and Mater. Soc., 8*, 30–34.
38. Rehmat, A., & Mensinger, M. C., (1996). Recovery of direct reduced iron from blast furnace dust. In: *Second International Symposium on Extraction and Processing for the Treatment and Minimization of Wastes* (pp. 27–30).
39. Gudenau, H. W., Denecke, H., & Wippermann, S., (1998). Iron ore and iron containing dust injection into the blast furnace. In: *2nd International Congress on the Science and Technology of Iron Making and 57th Iron Making Conference* (pp. 22–25).
40. Leguédois, V. O. S., Jongmans, F. T., & Chevallier, P., (2004). Morphology, chemistry, and distribution of neoformed spherulites in agricultural land affected by metallurgical point source pollution. *Environmental Pollution, 130*, 135–148.
41. Georgeta, P., Cornelia, P., & Gabriela, P., (1994). Petrographic assessment of coal and coke dusts in the vicinity of coke oven plants in Romania. In: Depers, A., & Bailey, J., (eds.), *White Paper and Abstracts. Working Group on Environmental Applications of Coal Petrology* (p. 29).
42. Cornelia, P., & Georgeta, P., (1999). *Atlas Carbopetrografic (Carbopetrographic Atlas).* Editura Academiei, Bucuresti.
43. Mihăiescu, D. C., Georgeta, P., & Cornelia, P., (2011). Evaluation of some blast furnace solid wastes composition by optical microscopy. In: Flores, D., & Marques, M., (eds.), *New Trends in Coal Science* (pp. 65, 66). 63rd Annual Meeting of the International Committee for Coal and Organic Petrology. Porto, Portugal. In: Memórias No. 18.

CHAPTER 3

A REFRIGERATION MODEL FOR STORAGE AND DISTRIBUTION SOLUTION OF COVID-19 VACCINES IN RURAL AND SEMI-URBAN AREAS OF INDIA

SURAJ BANDHEKAR,[1] RITESH KUMAR DEWANGAN,[2] PEEYUSH DEWANGAN,[1] RAM KRISHNA RATHORE,[1] TEJASVI ANANT,[1] and SHAIK IQBAL ALI BASHA[1]

[1]*Department of Mechanical Engineering, Rungta College of Engineering and Technology, Bhilai, Chhattisgarh, India*

[2]*Department of Mechanical Engineering, Rungta College of Engineering and Technology, Raipur, Chhattisgarh, India*

ABSTRACT

In India, the issue of invention, manufacturing, distribution, and storage poses a significant challenge, particularly in villages and small towns, due to the requirement of low temperature for the vaccine. There are two widely available and preferred types of vaccines, Covaxine and Covishield, both of which rely on chemical compositions that necessitate low temperature storage. The recommended temperature range for storing both vaccines is 2°C to 8°C. The most difficult stage lies in the distribution and storage of vaccines in rural and small town areas, where proper infrastructure and facilities are lacking. Additionally, a substantial amount of low temperature refrigeration is required. This chapter aims to assist in planning for improved storage and distribution of vaccines, with a reduced number of vial wastages.

Computational Optimization, Modeling, and Simulation for Engineering Applications.
Anupam Shukla, Sourabh Rungta, Mohan Awasthy, and Rakesh Himte (Eds.)

One potential solution is to collaborate with the Refrigeration Industries to develop and manufacture vaccine storage chambers.

3.1 INTRODUCTION

Almost all COVID vaccines need to be stored at temperatures below the surrounding atmospheric temperature. To ensure this, it is necessary to control and maintain the temperature of the transporting vehicle and storage facilities. Both Covaxin and Covishield vaccines contain chemicals that require very low temperatures for storage. Both vaccines need to be kept within a specific temperature range (between 2°C and 8°C). Bharat Biotech, the manufacturer of Covaxin, has developed a reusable package for transportation and storage that can maintain the vaccine at the required temperature for 28 days. However, the final stage of distribution poses significant challenges, particularly in rural areas and small towns where hospitals and health centers may lack the necessary infrastructure for storage and maintenance [3]. There is also immense pressure on equipment suppliers to provide large quantities of low-temperature refrigeration within a short timeframe. Additionally, there is limited available data to establish a comprehensive framework for transportation and storage. Therefore, there is a need for a quick, effective, and safe solution to address this issue faced by vaccine distribution authorities. Our study aims to propose a solution by utilizing existing resources such as refrigeration containers. By retrofitting these containers, they can be used to achieve the required storage temperatures [6].

3.2 EXPERIMENTAL METHODS AND MATERIALS

3.2.1 PROPOSED METHODOLOGY

The system will work on normal VCR (Vapor Compression Refrigeration), where we need to prepare a chamber or container working on the cycle can reduce and maintain the temperature to the desired low value, which is required for the Vaccines.

3.2.2 EXPERIMENTAL SETUP

A similar Experimental study has been done in U.S. which is being taken as our reference. The experimental setup for which is shown in the Figure 3.1. They used refrigeration container of the size 12.19 m × 2.44 m × 2.62 m this container is fully equipped to store the vaccines for the desired period.

Taking this as a reference, we can also develop such containers as per our requirements and atmospheric conditions.

The container must have the following:

- Thermocouples to measure the temperature at different locations in the chamber;
- Sensors for oxygen and carbon dioxide at proper places;
- Gauges to measure the weight of the dry ice being used for refrigeration;
- Also, the gauge to measure the temperature.

3.2.3 CONSTRUCTIONAL DETAILS

The suggested unit is made up of lightweight aluminum (or similar material) frame as the front or the outer wall for the container, which has the facility to be bolted at proper places. Inside, there is our refrigeration system containing all the components like compressor, evaporator, condenser fans for cooling. The system can operate on a 380/460 volt, three-phase supply with 50/60 hertz (Hz) power. The system also has a transformer which controls the power supply. The thermocouples can be replaced by digital temperature indicators and other electronic displays (Figure 3.1).

FIGURE 3.1 Sample system (container).

A local or the nearest dry ice manufacturing plant can be approached to prepare the boxes which are to be kept inside the container at various different segments. An empty small payload box can be fixed inside surrounded by the dry ice. To measure the temperature, a thermocouple can be fitted.

The test or running may have the following steps:

- Connect the unit to the 460 V 60 Hz, 3 phase power supply;
- Place the dry ice at the proper places in boxes;
- We need to place the boxes at the proper places in the container;
- Thermocouple connected gives the proper reading of the temperature time to time;
- Also check for oxygen and carbon dioxide levels should be about 20% and 0%, respectively;
- Fresh air ventilation should be 100%.

The components and the boxes containing the vaccine vials can be arranged as we can see in Figures 3.2 and 3.3, which show us a sample arrangement that we can follow or even modify. This arrangement can also be moved from place to place by attaching it to a big carriage vehicle.

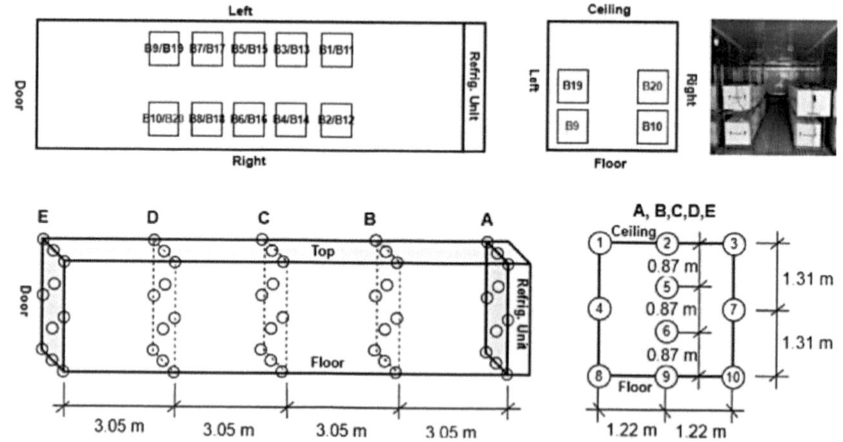

FIGURE 3.2 Sample arrangement of the container and the boxes.

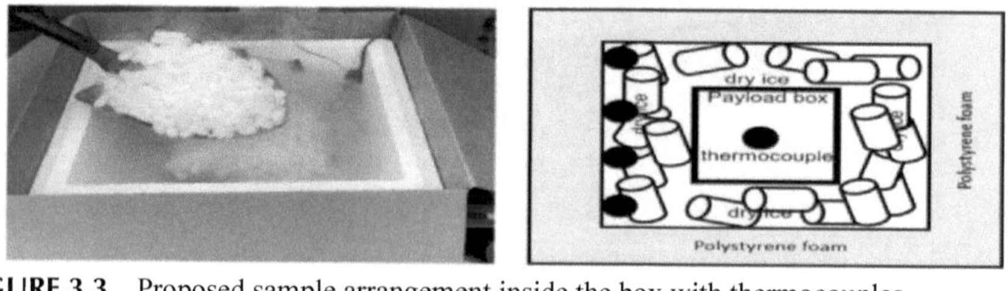

FIGURE 3.3 Proposed sample arrangement inside the box with thermocouples.

3.3 RESULTS AND DISCUSSION

Our aim is to reduce and maintain the temperature which is favorable for the vaccines. The arrangement proposed will give the desired results and outputs which include proper temperature distribution in the outer container, temperature distribution inside the dry ice package, sublimation of the dry ice, carbon dioxide content inside the chamber. As compared to the conventional storage system, this proposed system is able to give more effective results with proper accuracy. This enables the vaccines to be maintained at the desired temperature and conditions for a longer duration of time without the worry of being wasted or people waiting for too long.

3.4 CONCLUSION

By using this type of refrigeration system (container box) arrangement, the wastage of vials can be reduced and, in many cases, completely avoided. Also, the refrigerating system may help in keeping the vaccines safe for a longer span of time, so that they can be shifted to some other place where the requirement is more.

In the future, the study can be carried out with some alternate cycle or system of refrigeration, as we all know that these vaccines are to be taken every year (booster dose) even though 2 doses are complete, and still kids below 12 years of age are started with the vaccine dose. So, there is a lot of scope for the execution and improvement in the system.

KEYWORDS

- **carbon dioxide**
- **container box**
- **COVID-19**
- **refrigeration model**
- **semi-urban areas**
- **storage**
- **vaccines**
- **ventilation**

REFERENCES

1. Jian, S. M. Z., Anthony, G., Brian, F., Kashif, N., Kyle, G., Bo, S., Jeff, M., Joe, H., & Melissa, L., (2022). COVID 19 vaccine distribution solution to the last mile challenge: Experimental and simulation studies of ultra-low temperature refrigeration system. *International Journal of Refrigeration, 133*, 313–325.
2. Heilweil, R., (2020). *How Quick Can US Distribute a COVID-19 Vaccine? Here Are the Four Biggest Logistical Challenges*. Recode. https://www.vox.com/recode/22151473/vaccine-covid-19-pfizer-glass-syringes-needles-freezers (accessed on 21 December 2023).
3. Lindsey, R., (2020). *Climate Change: Atmospheric Carbon Dioxide*. Climate.gov.
4. https://www.climate.gov/news-features/understanding-climate/climate-change-atmospheric-carbon-dioxide (accessed on 21 December 2023).
5. Technol. ICOIACT, (2018). pp. 820–826. https://doi.org/10.1109/ICOIACT.2018.8350726.
6. Shinoda, T., &Budiyanto, M. A., (2016). Energy saving effect of roof shade for reefer container in marine container terminal. *J. Jpn. Inst. Navig., 134*, 103–113. https://doi.org/10.9749/jin.134.103.
7. Sun, J., Zhang, M., Fricke, B., Nawaz, K., Gluesenkamp, K., Shen, B., & Munk, J., (2021). *COVID-19 Vaccine Distribution Solution to the Last-Mile Challenge: Cryogenic Refrigeration to Facilitate the Maintenance of Ultralow Temperature*. ORNL/TM- 2021/1903.

CHAPTER 4

A BRIEF REVIEW ON QUETIAPINE DRUG CONTAINING EUTECTIC MIXTURE

RINKEE VERMA,[1] CHANCHAL DEEP KAUR,[2]
VIJENDRA SURYAVANSHI,[3] and RITIKA SINGH[1]

[1]*Assistant Professor, Rungta College of Pharmaceutical Science and Research, Raipur, Chhattisgarh, India*

[2]*Professor, Rungta College of Pharmaceutical Science and Research, Raipur, Chhattisgarh, India*

[3]*Assistant Professor, MJ College of Pharmacy, Bhilai, Chhattisgarh, India*

ABSTRACT

Eutectic mixture impacts significantly interested successful means of raising the dissolution rate and consequently the bioavailability is classified by low water-soluble drugs. Eutectic mixture of poorly water-soluble drugs with fatty acid as a carrier has reduced the incidence of these problems and enhanced dissolution. The point of this review chapter is on discrete methods of formation and explanation of the eutectic mixture. The different types of eutectic systems based on their little arrangement have been focused. Some of the practical features to be measured for the preparation of eutectic mixture, such as selection of carrier and methods of physicochemical characterization, have also been discussed. In this review, it is designed to discuss the future prospects related to the area of eutectic mixture manufacturing. Eutectic mixture, the amount of drug becomes compressed, or a crystalline pure drug is converted into an amorphous form, and hence the dissolution rate of the

Computational Optimization, Modeling, and Simulation for Engineering Applications.
Anupam Shukla, Sourabh Rungta, Mohan Awasthy, and Rakesh Himte (Eds.)

drug is increased. Evaluations of eutectic mixture are done by FTIR spectroscopy, XRD, SEM, DSC, solubility, and dissolution experiments.

4.1 INTRODUCTION

Numerous drugs are considered as Class II and IV drugs, as well as Quetiapine, a drug belonging to the class of antipsychotic drugs for the treatment of schizophrenia. In the biopharmaceutics classification system (BCS), which are very less soluble and have low permeability, rendering to Amid on GL, these drugs have a lower dissolution rate in aqueous medium, but once dissolved, they are easily absorbed in the GIT [1]. In this group, the dissolution of drugs is the rate-limiting step for absorption. The variation in the melting points between the two mechanisms determines the major component in the binary mixtures. Quetiapine has a unique receptor-binding profile, belonging to a new chemical class, the dibenzo thiazepine derivatives [2] (Figure 4.1). Quetiapine (quetiapine fumarate, Seroquel) is approved by the US Food and Drug Administration. Quetiapine is a BCS class II drug, having poor solubility and only 9% bioavailability [3].

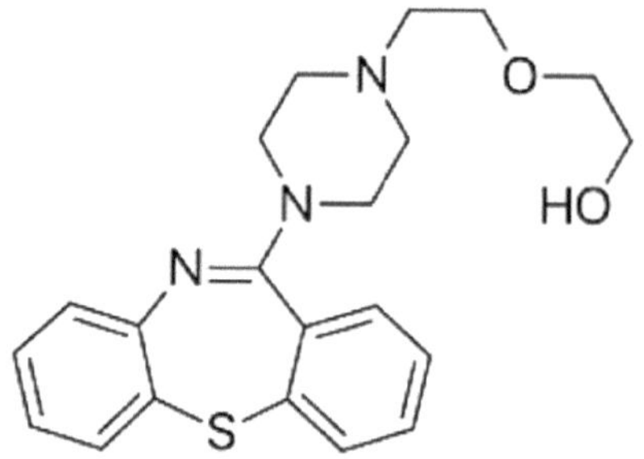

FIGURE 4.1 Structure of quetiapine.

1. **Absolute Bioavailability:** It measures the amount of active drug present in systemic circulation after oral, rectal, transdermal, and subcutaneous administration of the drug. Its purpose is used to identify a drug's innate absorption properties of an extravascular site.
2. **Relative Bioavailability:** This is a comparison of the same drug with different formulations. The most common and effective way to administer drugs orally is through ingestion. Many drugs face challenges in delivery for various reasons. Poor solubility of drugs due to modified drug absorption leads to low bioavailability [4].

4.2 REASON FOR POOR ORAL BIOAVAILABILITY

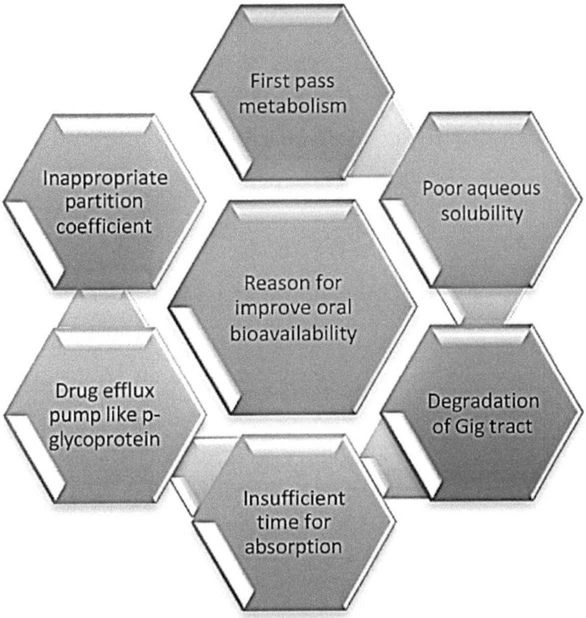

4.2.1 PRINCIPAL FACTOR OF BIOAVAILABILITY OF DRUGS

- Rate and amount of discharge of drug from various dosage forms;
- Consequent absorption from solution state;
- Bio-transformation method of absorption [5] (Table 4.1).

TABLE 4.1 Classification of Drug Using Biopharmaceutical Classification System (BCS)

Class I Drugs	**Class II Drugs**
High solubility	Low solubility
High permeability	High permeability
Ex: Propranolol, Metoprolol, Diltiazem, Verapamil	Ex: Ketoconazole, Mefenamic acid, Nifedipine, Nircardipine
Class III Drugs	Class IV Drugs
High solubility	Low solubility
Low permeability	Low permeability
Ex: Acyclovir, Neomycin, Captopril, Alendronate	Ex: Chloro-thiazide, Furosemide, Tobramycin

4.2.2 NOVEL DRUG DELIVERY SYSTEM

Nano-suspensions, micro-emulsion DDS, solid lipid nanoparticles (SLN), polymeric nanoparticles, Vesicular transport systems such as lysosome, liposome, etc. Solid dispersion (SD) techniques are the most important parameters for understanding the physicochemical properties of the drug. The preparation method and amount of carrier play an important role in enhancement of drug dissolution rate [7, 8].

4.3 SOLID DISPERSION (SD)

4.3.1 DEFINITION

The employment of solid dispersion (SD) to achieve dissolution rate of oral absorption of low water-soluble drugs. Also recommended the regulation of a eutectic mixture of a hydrophobic drug with a physically impotent, easily miscible carrier [9]. The system is either crystalline or amorphous form. Mostly amorphous form is easily solubilized than the crystalline form because no potency is required to split up the crystal lattice of drug during dissolution process [10] (Table 4.2).

TABLE 4.2 Classification of Different Types of Solid Dispersion

S. No.	Solid dispersion type	Matrix	Drug	Conclusion	Phase
1	Eutectics	C	C	First type of solid dispersion	2
2	Amorphous precipitation in crystalline matrix	C	A	Rarely Avoidance	2
3	Solid solutions	-	-	-	-
A	Continuous solid solution	C	M	Miscible at all composition, never prepared	1
B	Discontinuous solid solution	C	M	Partially miscible two phases even though drug is molecularly dispersed	2
C	Substitution solid solution	C	M	Molecular diameter of solute (drug) liters less than 15% from the matrix system. When the drug is substitution than the drug and matrix system are continuous or discontinuous. In case discontinuous 2 phases of drug are molecularly dispersed.	1 or 2
4	Intestinal solid dispersion	C	M	Drug having molecular diameter less than 59% of matrix diameter. Mostly limited miscibility, discontinuous	2
5.	Glass suspension	A	C	Particle size of dispersed phase obtained after crystallization of drug	2

4.3.2 TYPE OF SOLID DISPERSION (SD)

4.3.2.1 SIMPLE EUTECTIC MIXTURE

A simple eutectic mixture is considered as a ratio of two or more component which are completely soluble in liquid state but only a very controlled proportion in the solid state (Figure 4.2). When mixture A and B combination with E is cooled, A and B crystallize out together [8, 9].

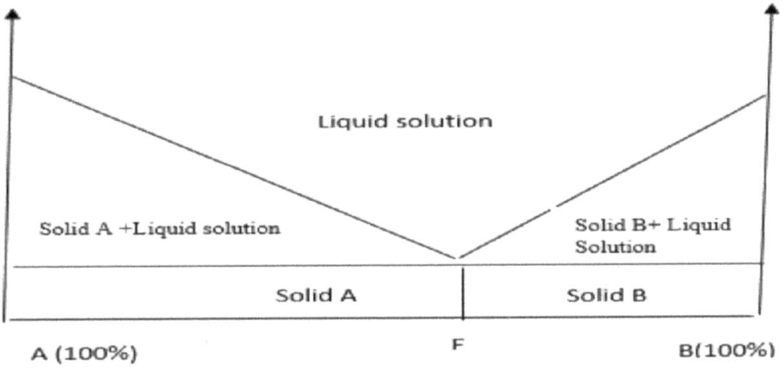

FIGURE 4.2 Simple eutectic mixture.

4.3.2.2 SOLID SOLUTION

The solid solution consists of a solute soluble in a solvent. A mixture of a mixed ratio of crystal shapes because of two ratios of elements crystallizing each other in the same component form a one-phase system. Solid solutions can be classified according to their solubility into two types: continuous and discontinuous solid solutions.

4.3.2.3 CONTINUOUS SOLID DISPERSION (SD)

The component of mixture soluble in all ratios which shows that the property of bonding strength in between two ratios is stronger than the bonding strength of molecules of other components.

4.3.2.4 SUBSTITUTION CRYSTALLINE SOLID SOLUTION

The solid solution having crystalline formation, in solute molecule become displace for solvent molecule in the crystal arrangement or apply into the space in fundamental between solvent molecule [8, 9] (Figures 4.3 and 4.4).

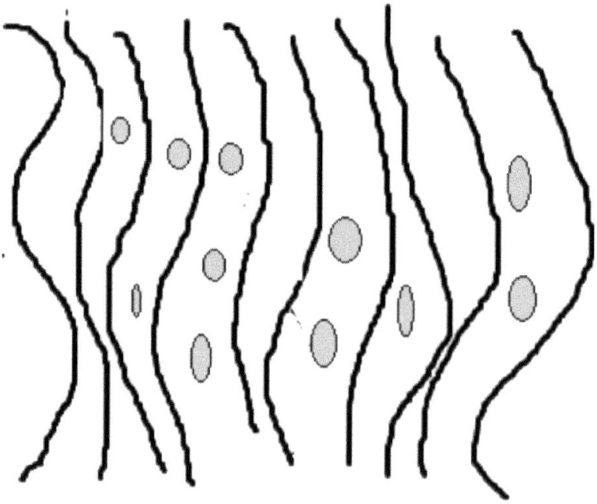

FIGURE 4.3 Interstitial crystalline solid solution.

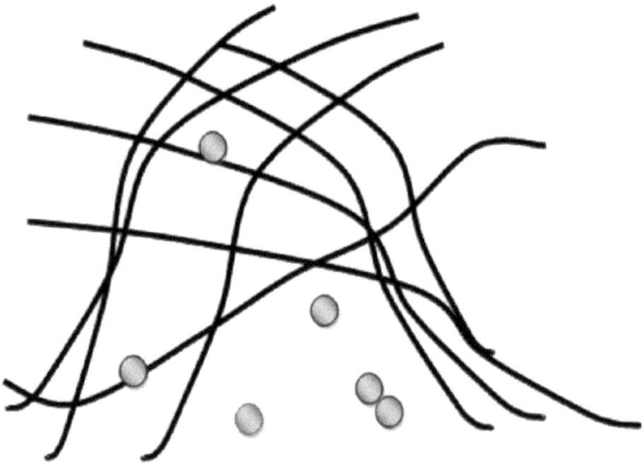

FIGURE 4.4 Amorphous solid solution.

4.4 MECHANISM OF INCREASED DISSOLUTION RATE

The improvement of dissolution rate for SD can be credited to a number of factors:

1. **Reduction of Particle Size:** Solid solution, amorphous dispersions, and particle size are compacted to a minute level. This can show the improvement of dissolution rate due to an increase in the surface area solubilization.

2. **Solubilization Effect:** The carrier material (CM), as it dissolves, may have a solubilization effect on the drug.

4.5 MANUFACTURING METHOD OF SOLID DISPERSION (SD)

4.5.1 FUSION METHOD

The carrier ratio material is heated to a given temperature just above its melting point and the drug is fused into the medium. The mixture is cooled with continuous stirring to equally disperse the drug throughout the medium. Particle size decreases under this process, fusing the method to the final level, leading to molecular dispersion of the drug in the carrier matrix (Figure 4.5).

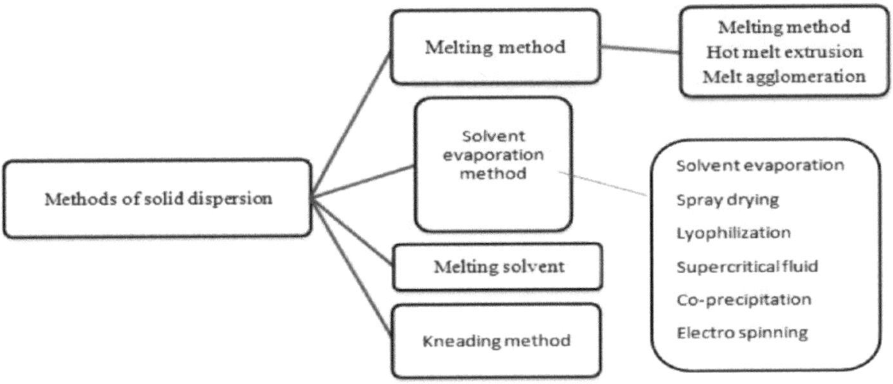

FIGURE 4.5 Method of manufacturing.

4.5.2 SOLVENT METHOD (SM)

Solvent method (SM) the active pharmaceutical ingredient (API) and CM are solubilizing in a suitable organic solvent. Then the solvent is vaporized at a higher temperature. As the solvent is being removed. The co-precipitate is then dried under vacuum conditions.

4.5.2.1 SOLVENT SELECTION

It is very difficult to choose a solvent. Usually, CM is hydrophilic and the active form of the drug is lipoidal. Drug selection of a common solvent is difficult and its complete removal, required by its toxic nature, is authoritative. Solvents may plasticize polymeric carriers, e.g., polyvinylpyrrolidone (PVP), making their complete removal even more difficult.

4.6 CHARACTERIZATION

4.6.1 CONTACT ANGLE

The principle of contact angle is to measure the surface free energy which gives a best approach to determining the properties of solid. The surface energy of a solid can be observed from the determination of contact angle of pure liquid drop on the solid surface. Main point is to determine the surface energy, wetability (Figure 4.6).

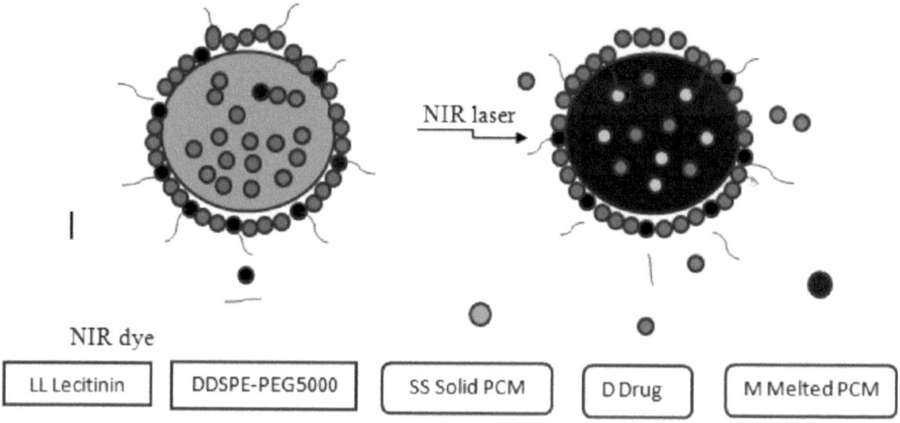

FIGURE 4.6 Surface energy of a solid dispersion.

KEYWORDS

- **bioavailability**
- **eutectic mixture**
- **physicochemical properties**
- **quetiapine**
- **solid dispersion**
- **solubility**
- **surface energy**

REFERENCES

1. Lipinski, C. A., Lombardo, F., Dominy, B. W., & Feeney, P. J., (2001). Experimental and computational approaches to estimate solubility and permeability in drug discovery and development settings. *Adv. Drug Deliv. Rev., 46*, 3–26.
2. Lobenberg, R., & Amidon, G. L., (2000). Modern bioavailability, bioequivalence, and biopharmaceutics classification system. New scientific approaches to international regulatory standards. *Eur. J. Pharm. Biopharm., 50*, 3–12.
3. Amidon, G. L., Lennernas, H., Shah, V. P., & Crison, J. R., (1995). A theoretical basis for a biopharmaceutics drug classification: The correlation of *in vitro* drug product dissolution and *in vivo* bioavailability. *Pharm. Res., 12*, 413–420.
4. Curatolo, W., (1998). Physical chemical properties of oral drug candidates in the discovery and exploratory development settings. *Pharm. Sci. Technol. Today, 1*, 387–393.
5. Granero, G. E., Ramachandran, C., & Amidon, G. L., (2005). Dissolution and solubility behavior of fenofibrate in sodium lauryl sulfate solutions. *Drug Dev. Ind. Pharm., 31*, 917–922.
6. Singhal, D., & Curatolo, W., (2004). Drug polymorphism and dosage form design: A practical perspective. *Adv. Drug Deliv. Rev., 56*, 335–347.
7. Law, D., Wang, W., Schmitt, E. A., et al., (2003). Properties of rapidly dissolving eutectic mixtures of poly (ethylene glycol) and fenofibrate: The eutectic microstructure. *J. Pharm. Sci., 92*, 505–515.
8. Kreuter, J., (1999). Feste dispersionen. In: Kreuter, J., & Herzfeldt, C. D., (eds.), *Grundlagen Der Arzneiformenlehre Galenik* (Vol. 2, pp. 262–274). Springer, Frankfurt am Main.
9. Shargel, L., & Yu, A. B., (1999). *Applied Biopharmaceutics & Pharmacokinetics* (4th edn.) New York: McGraw-Hill. ISBN 0-8385-0278-4.
10. Hite, M., Frderici, C., & Turner, S., (2003). Part 1: Oral delivery of poorly soluble drugs. *Pharmaceutical Manufacturing and Packaing Sourcer* (No. 1–3).
11. Leuner, C., & Dressman, J., (2000). Improving drug solubility for oral delivery using solid dispersion. *Eur. J. Pharm. Biopharm., 50*, 47–60.

CHAPTER 5

A REVIEW ON PHYTOPHARMACOLOGICAL PROFILE OF EMBELIA RIBES

ROSHNI SHARMA and SHREYA SINGH

Faculty, Rungta College of Pharmaceutical Science and Research, Raipur, Chhattisgarh, India

ABSTRACT

Embelia ribes is also called false black pepper be a member to the family of Primulaceae. It is a forested bush, should be recorded in Ayurveda for use to cure numerous disorders, like diabetes. It's globally disseminated in many countries like India and other Asian countries. Great functional components from the entire portion of ER are embelin, alkyl substituted hydroxy benzoquinone. It's a naturally obtained substance. *ER* Burm., referred to as the former age, is present as '*Chitrmandula*' or '*Krimighna*' form. ER is the chief element in Ayurveda for making many types of Ayurvedic formulations. The organoleptic characters of *ER* Burm. f. are that there are various stomata present in its bark. Its leaves have a leathery or rough surface, 5–9 by 2–3.8 cm. ER is elliptic or oblong in shape and its leaves have a smooth surface. Three new compounds identified as embelinol, embeliaribyl ester, and embeliol were isolated through *ER* along with embelin. There are several pharmacological actions on humans like analgesic activity, antianxiety activity, antibacterial activity, etc., it's an acceptable substance for the therapy of central nervous system (CNS) disease and chronic diseases like cancer. There are numerous *in-vitro* and *in-vivo* animal model studies that have been done with the help of ER. Embelin has many pharmacological

Computational Optimization, Modeling, and Simulation for Engineering Applications.
Anupam Shukla, Sourabh Rungta, Mohan Awasthy, and Rakesh Himte (Eds.)

uses in ayurvedic formulations like Krumikushta; it is useful in skin diseases or numerous traditional uses. There have been several recent developments, synthesis, and biological applications in embelin that have been done.

5.1 INTRODUCTION

Embeliaribes (ER) Burm. (Family: Primulaceae), it is a bush, and ER is mentioned in Ayurvedic medicine for the treatment of several disorders, like diabetes, and some other diseases like treating joint diseases and some diseases of the skin [1]. It is widely distributed in India and many other Asian countries [1]. It is referred to as the earliest times, in the pattern of the drug '*Chitrmandula*' or '*Krimighna*' should be chief constituents in various Ayurvedic preparations. Embelin is the chief active component having many different parts of ER [2]. Embelin is present in the whole plants of *Embelia ribes*. Different parts of *embelia* have different amounts of embelin. The active component of ER was first isolated half a century ago. The tincture of ER fruits is scientifically recognized in the earliest times for its insulin reactions activity in glucose-fed Florida white bunny [2]. A recent study states that there are byproducts of embelin that may be used in the treatment and prevention of obesity-linked type-2 DM, etc. [3].

5.2 PHARMACOGNOSY OF *EMBELIA*

5.2.1 MORPHOLOGY

The organoleptic characters of *ER* Burm. f. are that there are various stomata present in its bark. Its leaves have a leathery or rough surface, measuring 5–9 by 2–3.8 cm. ER is elliptic or oblong in shape and its leaves have a smooth surface. ER flowers are small in size and greenish in color. Its sepals are triangular in shape and cilia are present on them. *Embelia* flowers have 5 petals, and its flowering time is February. Fruits of *embelia* are dark black and a little bit red in color, with a diameter of 2.4–4.0 cm and a watery and smooth surface.

5.2.2 ANATOMY OF STEM

The mature stem T.S. is rather disc-like outline. Outer layer of stem are single layered and epidermis cells are present. In it multicellular glandular trichomes are also present. In parenchymatous region tannins cells are more prominent [4].

5.2.3 ANATOMY OF ROOT

The mature root T.S. is somewhat ring-shaped outline. Outer layer of stem are single layered and epidermis cells are present. In it multicellular glandular trichomes are also present. Root cork having 15–17 layered schizogenous cavities. Secondary types of vascular tissues are found underneath this cavity. Tannin cells are tremendously conspicuous with biseriate medullary rays [4].

5.2.4 ANATOMY OF LEAF

I In these leaves, T.S. monocotyledon characters can be easily visible. In the leaf of *embelia*, a single layer of epidermis cells is present and there is no filamentous outgrowth present on it. In the leaf of *embelia*, the mesophyll comprises a single layered palisade and generous intercellular spaces with spongy tissue. Plenty of cells are copious in tannin. Close by the midrib, oil glands are overly usual in the mesophyll. 'C' shaped wide vascular bundles are present in it. Phloem fibers are very prominent, and stomata are of Ranunculaceae's type [4].

5.2.5 ISOLATION OF EMBELIA

Embelinol, embeliaribyl ester, and embeliol are 3 new elements that are recognized and excreted out in the seeds of *ER*, accompanied by embelin. Potassium, chromium, copper, calcium, zinc, and manganese are available in the seeds of ER. An uncommon nitrogen-containing various types of derivatives are excreted out in the roots of *ER* using ethanolic extract [5].

For the determination of embelin through the *ER*, a HPLC method was devised. A quantity of Embelin in ER of 4.33% w/w was detected. TLC is the suggested method that can be used for the quantitative ascertainment of embelin in *Embelia* plants [6].

5.3 PHARMACOLOGY OF *EMBELIA*

5.3.1 PALLIATIVE ACTIVITY

E. ribes was deliberate as a powerful centrally acting painkiller. In the rat brain, *in vivo* studies were carried out to investigate the binding site of embelate of K. In the brain, the main binding sites may be involved in the analgesic action [7].

5.3.2 ANTIBACTERIAL ACTIVITY

Embelin exhibits antibacterial activity against staphylococci organisms and inhibits the growth and development of pseudomonas types of organisms[8].

5.3.3 ANTIFERTILITY ACTIVITY

As stated by the research, everyday administered *Vinca rosea* extract with the embelin in the male albino rats caused an abrupt increase in the amount of alkaline phosphatases and acid of testis and prostate, specifically refining the digestive function. This gives rise to the study for its anticipated activity in engendering antifertility [9].

5.3.4 ANTIDIABETIC ACTIVITY

*E. ribes*berries extracts administered through the mouth it is causing bring down blood glucose level in by the range of around 13% to 20% normo-glycemic ZF rats and around 28.1 to 28.2% and approximately 34.5% in alloxan actuated diabetic affected rats. Consequently, *E. ribes* berries were proposed to hold active hypoglycemic components, and its antidiabetic activity perhaps shown caused by its direct action on tissues and expected to expand secretion of insulin. To synchronize the insulin resistance, embelin was assessed. Embelin lowers the increased weight, blood glucose level, and insulin in plasma of diabetic rats [10].

5.4 ROLE OF EMBELIN IN INCURABLE DISORDERS

5.4.1 ANTICANCER ACTIVITY OF EMBELIN

Embelin has various kinds of anticancer effects, such as breast cancer, prostate cancer (PC), hepatic cancer, pancreatic cancer, colon cancer, gastric cancer, leukemia, and various myeloma. These effects have been tested on living subjects, including animals, plants, and whole cells, both inside and outside of a living organism. It can effectively inhibit various types of tumorous cell accumulation and movement, induce tumor cell death, and hinder tumor growth and development. Embelin can induce death of neoplasm cells by downregulating the expression of XIAP and activating the caspase cascades. It also blocks various signaling pathways and has been found to modulate the expression of various gene products, such as antiapoptotic and metastatic factors, by inhibiting NF-jB. Embelin also inhibits the expression of COX-2, MMP-9, cyclin D1, ICAM-1, c-Myc, and VEGF, all of which are controlled by NF-jB and closely associated with tumor growth, invasion, and angiogenesis.

5.4.2 EMBELIN AND AUTOIMMUNE INFLAMMATORY ILLNESS

Crohn's disease, inflammation of joints, red patches with white scales on top, *Encephalomyelitis disseminata,* and acute disseminated encephalomyelitis (ADEM) are encompassed in various types of diseases like redness, swelling, heat, fever, and pain. They have their own usual diagnostic attributes, namely unknown cytology, immune connection, and inadequate effective curative components with a lack of intricacy (Figure 5.1).

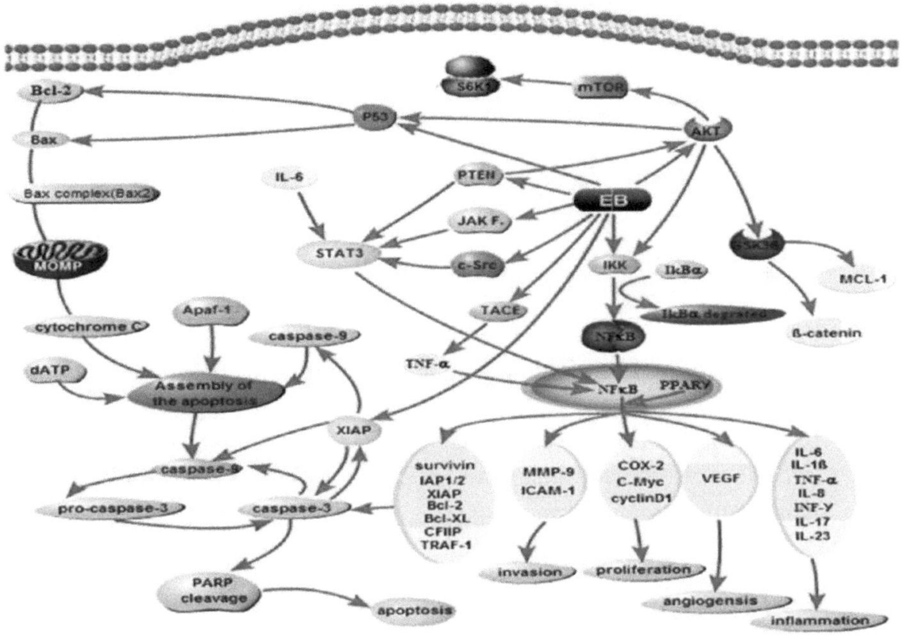

FIGURE 5.1 Graphic representation of antitumor mechanism of ER. [EB: Embelin; XIAP: inhibitor of X-chromosome linked cell death protein; NF-jB: nuclear factor-jB; IjBa: the IKK IjB kinase, constrained of NF-jB, AKT protein kinase B; GSK3b: glycogen synthase kinase-3b PPARc peroxisome proliferator-activated receptor gamma; JAK: Janus-like kinase; mTOR: mammalian target of rapamycin; S6K1: ribosomal S6 kinase 1; and STAT3: signal transducer and activator of transcription 3].

5.5 PHARMACOLOGICAL USES OF *EMBELIA*

5.5.1 AYURVEDIC FORMULATIONS

ER also used customarily in numerous health care motives and ayurvedic preparations [11]. Some of the commonly used formulations are given in Table 5.1.

TABLE 5.1 Ayurvedic Formulation of *Embeliaribes*

Formulation	Uses
Sara	Helps in movements of components in intestinal tract.
Krumihara	Used in to cure worm infestation.
Krumikushta	Used in to cure various infective skin diseases.
Pramehahara	To treat diabetes.
Shirorogahara	To treat headache.
Vahnikara	Helps to improve improves appetite.
Shulahara	To treat abdominal colic pain.

5.5.2 CONVENTIONAL USES

ER has been conventionally used in numerous healthcare motives for over 5,000 years [12]. The conventional uses like:

- It detoxes blood; hence it is useful in an extensive range of skin disorders;
- Along with long pepper embelin shows contraceptive effect;
- Improves skin complexion and rejuvenation purposes, embelin is used;
- It has lenient diuretic action;
- In contrast to bloating, vomiting, indigestion, constipation, gastritis are cured by embelin.

5.6 THE RECENT DEVELOPMENTS OF EMBELIN AND IN SYNTHESIS OR ITS BIOLOGICAL APPLICATIONS

5.6.1 REACTIVITY OF EMBELIN

The hydroxyl groups attending at the C2 and C5 positions are accountable for formation of numerous metal compounds and for exchange by several reactions such as esterification, sulfonylations, and the biological activities of embelin hydrophobic alkyl side chain plays a major role [13].

5.6.2 REACTIONS AT C1 AND C2 OF EMBELIN

These responses involve the preparation of several metal compounds. Switch to benzoyl ester derivatives of embelin has been made using several benzoyl halides and modify the OH group at C2 to that of OMe, OAc groups.

5.6.3 REACTIONS AT C4 AND C5 OF EMBELIN

Comfortably attainable reactions like dimerization, conjecture, and synthesis of oxadiazines, phenazines were carried out at the carbonyl group at C4 and the hydroxyl group at C5 of embelin [14].

KEYWORDS

- **Antibacterial activity**
- **Anticancer activity**
- **Autoimmune inflammatory illness**
- **Biological applications**
- **Coriaceous elliptic**
- **Embelin**
- **Shrub ancient lenticels**

REFERENCES

1. Guha, B. D. N., Sensarma, P., Pal, D. C., & Lexicon, A., (2001). *Medicinal Plants of India* (p. 135). Naya Prakashan, Calcutta, India.
2. Tripathi, S. N., (1979). Screening of hypoglycemic action in certain indigenous drugs. *J. Indian Med. Yoga Homeopath, 14,* 159–169.
3. Stalin, A., Irudayaraj, S. S., Gandhi, G. R., Balakrishna, K., Ignacimuthu, S., Al- Dhabi, N. A., (2016). Hypoglycemic activity of 6-bromoembelin and vilangin in high-fat diet fed-streptozotocin-induced type 2 diabetic rats and molecular docking studies. *Life Sci., 153,* 100–117.

4. Warrier, P. K., Nambiar, V. P. K., & Ganapathy, P. M., (2001). *Some Important Medicinal Plants of the Western Ghats, India: A Profile* (pp. 139–156). International Development Research Centre (IDRC), Canada.
5. Lin, P., Li, S., Wang, S., Yang, Y., & Shi, J., (2006). A nitrogen-containing 3-alkyl-1,4-benzoquinone and a gomphilactone derivative from *Embelia ribes*. *J. Nat. Prod., 69*(11), 1629–1632.
6. Sudhakar, R. S., Unnikrishnan, K. P., Ravindran, P. N., & Balachandran, I., (2005). Determination of embelin in *Embelia ribes* and *Embelia tsjeriam-cottam* by HPLC. *Indian Journal of Pharmaceutical Sciences, 67*(6), 734–736.
7. Zutshi, U., Johri, R. K., & Atal, C. K., (1989). Possible interaction of potassium embelate, a putative analgesic agent, with opiate receptors. *Indian Journal of Experimental Biology, 27*(7), 656, 657.
8. Mohammad, A. K., Menda, A. N., & Zafar, A., (2010). *In vitro* antimicrobial activity of fruits extract of *Embelia ribes* Burm. *International Journal of Pharmaceutical & Biological Archives, 1*(3), 267–270.
9. Chauhan, S., Agrawal, S., Mathur, R., & Gupta, R. K., (1979). Phosphatase activity in testis and prostate of rats treated with embelin and *Vinca rosea* extract. *Experientia, 35*(9), 1183–1185.
10. Uma, B., Raman, K., & Pillai, K. K., (2000). Effect of ethanolic extract of *Embelia ribes* on dyslipidemia in diabetic rats. *Int. Jnl. Experimental Diab. Res., 3*, 159–162.
11. Hu, R., Zhu, K., Li, Y., Yao, K., Zhang, R., Wang, H., Yang, W., & Liu, Z., (2011). Embelin induces apoptosis through down-regulation of XIAP in human leukemia cells. *Med. Oncol., 28*, 1584–1588.
12. Chen, J., Nikolovska-Coleska, Z., Wang, G., Qiu, S., & Wang, S., (2006). Design, synthesis, and characterization of new embelin derivatives as potent inhibitors of X-linked inhibitor of apoptosis protein. *Bioorg. Med. Chem. Lett., 16*, 5805–5808. https://doi.org/10.1016/j bmcl.2006.08.072.
13. Anonymous, (2001). *The Ayurvedic Pharmacopoeia of India* (Vol. 1, No. 1, pp. 163–165). Govt of India, Ministry of Health and Family Welfare, Dept. of Aysuh.
14. Shaikh, A., Dhadde, S. B., Durg, S., Veerapur, V. P., Badami, S., Thippeswamy, B. S., et al., (2016). Effect of embelin against lipopolysaccharide-induced sickness behavior in mice. *Phytother. Res., 30*, 815–822. doi: 10.1002/ptr.5585.

CHAPTER 6

IDENTIFICATION OF TOMATO LEAF DISEASE PREDICTION USING CNN

SURAJ KUSHWAHA[1] and SHRIKANT ZADE[2]

[1]*Department of Computer Science and Engineering, OIST, Bhopal, Madhya Pradesh, India*

[2]*Faculty of Computer Science and Engineering, OIST, Bhopal, Madhya Pradesh, India*

ABSTRACT

Tomatoes are the large-scale planted vegetable crop in India's lands of agriculture. Even though, the tropical climate is beneficial for its growth, specific environment and other variables influence tomato plants' growth. In addition to, these environmental conditions and naturalistic disasters, plant disease is a severe agricultural production issues that outgrowth in economic loss. Therefore, quick illness detection can provide a better end result than current detection algorithms. As a result, the deep learning outlooks based on computer vision might be used to explore diseases too soon. This study thoroughly looks at the disease categorization and detection approaches used to recognize tomato leaf diseases. The availability and limitations of the strategies provided are also discussed in this study. In the end, employing a hybrid deep-learning architecture, this research produces an anonymous disease detection approach for detecting tomato leaf disease.

6.1 INTRODUCTION

Agriculture is very much more than just a source of food for ever-increasing populations. Plant diseases, on the other hand, are threatening the resource's

Computational Optimization, Modeling, and Simulation for Engineering Applications.
Anupam Shukla, Sourabh Rungta, Mohan Awasthy, and Rakesh Himte (Eds.)

survival. In addition, plant diseases have caused significant economic losses in agriculture and forestry. As a result, early detection and prognosis of plant diseases are critical for taking prompt measures. We can detect illness in plants using a variety of approaches. Some ailments, on the other hand, are difficult to diagnose early on. They'll have to wait for a bit before they can find out what's going on. Under these circumstances, advanced analysis, which is generally done with robust microscopes, is required. Diseases harm the health of the plant, which has an impact on its growth. Therefore, it is critical to monitor the progress of the farmed crop to guarantee minimal losses. Unfortunately, there are a slew of tomato illnesses that are wreaking havoc on the crop's leaves. The main goal of the proposed study is to develop a solution to the issue of detecting tomato leaf disease using the most straightforward approach while using the fewest computational resources possible to obtain results equivalent to state-of-the-art approaches. Automatic feature extraction is used to help categorize input images into illness classifications. The suggested system gained an average purity of 94%–98%, demonstrating the neural network outlook's viability even under adverse conditions.

This chapter uses a miniature version of the convolutional neural network (CNN) model to detect and diagnose illnesses in tomato leaves (Figure 6.1).

FIGURE 6.1 Infected leaf.

In other situations, the indications can merely be detected in non-visible areas of the electromagnetic spectrum. This study aims to create a user-friendly system for farmers that will assist them in identifying tomato plant problems without having to call in an expert. We first get an image from the user and then extract features from it. We use photo conversion and scaling to extract the characteristics. We will use CNNs to classify illnesses. The system may use a single picture database, with 80% of the data being used to train the model and 20% testing the model.

6.1.1 SYMPTOMS OF TOMATO LEAF DISEASE

As the plant responds to the disease, it may show subtle color, shape, or function changes. Here, we'll go through the signs of these illnesses that we should be aware of if our plant's development seems to be slowing down. The key to preventing agricultural losses is proper categorization and identification of leaf diseases. Different plant leaves carry various illnesses and display various symptoms.

6.1.2 LEAF SPOT ON SEPTORIA

One of the major prevalent tomato plant's leaf diseases is the Septoria leaf spot. This fungus causes a tiny, round area with a grayish-white center and black margins, the first sign of its presence. Tiny black flecks may emerge in the middle. The leaves of the affected tomato plants become yellow, wither, and fall off—long stretches of hot, humid weather.

6.1.3 EARLY BLIGHT (ALTERNARIA)

Alternaria (early blight) is a fungus that causes leaf spot or early blight on tomatoes. Brown or black patches with dark borders appear on the lower leaves, almost like a target. Fruit stem ends are susceptible to assault, resulting in enormous, sunken black spots with concentric rings. The fungus that causes this tomato plant disease generally emerges after the plants have produced fruit.

6.1.4 BLIGHT IN THE LATE STAGES

Late blight is a tomato plant disease caused by the fungus Phytophthora infestans. It develops during chilly, wet conditions after the growth climate. It appears on foliage as frost damage, with irregular green-black splotches. Large, irregular-shaped dark patches on fruits can swiftly destroy them.

This fungus that causes tomato plant disease also affects potatoes and can be spread from them. The same restrictions as for Septoria leaf spot should be used.

6.2 REVIEW OF LITERATURE

For the growth of the leaf health recognition, for the task of tomato plants, we used data from Kaggle Tomato leaf disease detection dataset. A leaf image of a tomato plant was used to build the dataset, including healthy and ill shots with different classifications. We will split the data into sections: we will use 80% of the dataset to train the neural network, and another 20% will be utilized as the test set. In the next stage, we'll utilize the set of photographs to train the model. So far, we've been successful in identifying the diseased leaf. In the future, we will classify diseases based on whether they are caused by viruses, fungus, or bacteria and provide the farmer in the field instructions on how to treat them. Based on the study of numerous authors' work, the following information regarding the tomato plant disease.

Khirde et al. [1] reviewed different segmentation and feature extraction algorithms that may be used to diagnose plant illnesses using images of their leaves in their literature survey. Manually detecting plant illnesses is extremely tough due to time, understanding of plant diseases, and labor. Therefore, the complete procedure of detecting plant leaf diseases has been separated into five phases by the author, which is as follows: (i) image capture; (ii) pre-processing; (iii) segmentation; (iv) feature extraction; and (v) disease categorization. For RGB leaf image capture, the transformation structure was utilized. The picture is then pre-processed to reduce noise and improve contrast. Next, segmentation separates an image into multiple feature parts using K-means clustering, Otsu filters, and other techniques. This segmented picture is then utilized to extract features before being classified using a variety of classification techniques.

Sannakki et al. [2] developed a feed-forward backpropagation Neural Network-based approach to identify and categorize illnesses in grape leaves. For the diagnosis, the author used photos of a grape leaf with a complicated backdrop. Anisotropic diffusion is then utilized to eliminate the image's noise before it is segmented using k-means clustering. Finally, neural networks are used to observe the outcomes. The results were tested on downy mildew and powdery mildew matrices, with actual positive and false-positive para-author claims to have an accuracy of about 100% when using the color characteristic alone.

Kutty et al. [3] classified the downy mildew and anthracnose watermelon leaf diseases using a neural network-based technique. To assess the efficiency of the suggested approach, the authors calculated the valid positive rate, actual negative rate, and total accuracy. This classification is based on the RGB color model's extraction of color features from identified pixels in the region of interest.

Arjun et al. [4] The study's goal is to offer a foundation for monitoring and assessment by presenting critical data regarding the agricultural scenario and its contribution to total GDP.

In this research study, Shrivastava et al. [5] presented practical applications of digital image processing in agriculture for identifying and categorizing Brown Spot and frog eye. They employ digital image processing to extract the form feature vector, which is then used to classify diseases using a K-NN classifier.

Zhang et al. [6] provide a new technique based on global features and zone-based local features for detecting citrus canker from field leaf photos, which is more challenging than lab leaf images. To begin, an updated Adaboost algorithm is utilized to choose the most critical aspects of citrus lesions for segmentation from the background.

Jadhav et al. [7] introduce a pre-trained CNN-based soybean disease detection system based on transfer learning such as AlexNet, GoogleNet, VGG16, ResNet101, and DenseNet201.

Dasmunshi et al. [8] this study aims to monitor plant health using the NDVI (Normalized Difference Vegetation Index) computation, which can help distinguish between healthy and unhealthy plants by calculating their NDVI values. The photos of the plant were captured using a NIR camera connected to the Raspberry Pi.

Patil et al. [9] Traditional procedures were ineffective and incorrect. As a result of several studies, image processing has been included for reliable disease diagnosis utilizing plant leaves. The disease may be detected by looking for different spots and patterns on plant leaves. The adoption of digital image processing for more accurate findings was a step forward. However, after consulting several reputable IEEE, international conferences, and international journal publications in this sector, it was discovered that none of them provides a cure for the plant ailment.

Mhatre et al. [10] used the leaf recognition method with the feature recognition system. The CNN is used to identify plant leaves. PCA extracts and improves properties before incorporating them into CNN.

6.3 LEAF DISEASE DETECTION AND CLASSIFICATION

Deep learning is a subpart of the machine learning techniques. It consists of various techniques such as CNN, RNN, ANN, LSTM. Here, is a general block diagram of CNN performance of classification on image (Figure 6.2).

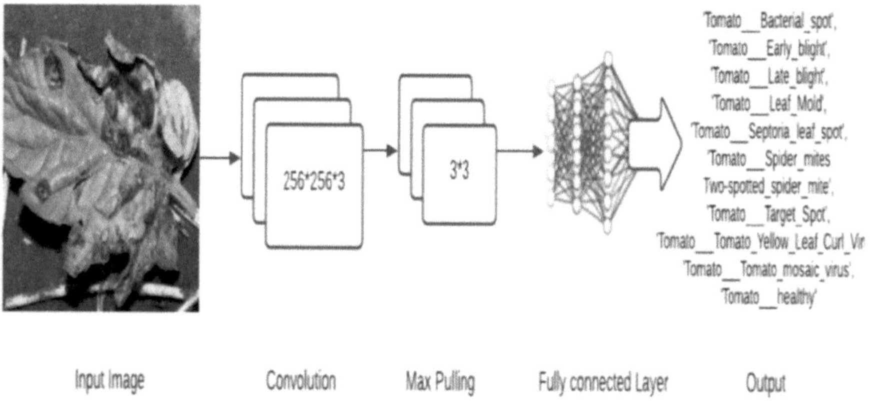

FIGURE 6.2 General block diagram of CNN approach.

Images are divided into two parts, first is training and second is testing. The ratio of training and testing is 80–20, so 80% of the image data is used for training the CNN model and the remaining 20% of the data will be used for testing.

Here is the convolution layer that helps to extract the features of the input images. The process of feature extraction is like this: we can extract multiple features from a single input image. In our case, we extract 32 features in a single input image.

Then we apply a Max pool layer that helps to extract maximum features from a given frame. After that, we apply a flatten layer to the given input image, and a flatten layer is used to pull feature maps to a single one-dimensional vector array.

That is fed into a fully connected layer with 1,568 neurons in the hidden layer and 10 neurons in the output layer. At the end, the overall accuracy score will be used for permutation evaluation (Figure 6.3).

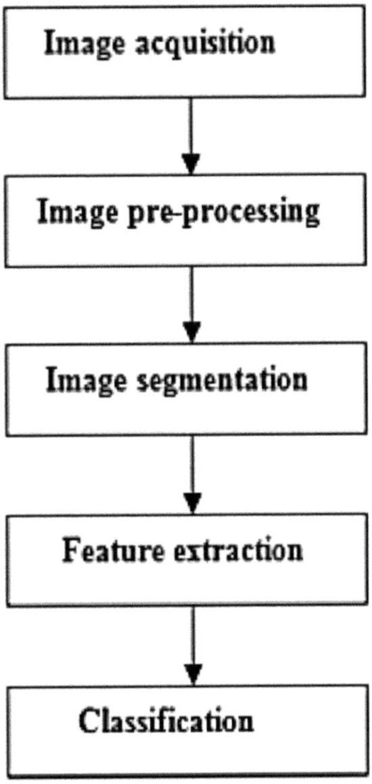

FIGURE 6.3 General block diagram for feature extraction.

Based on feature extraction, the classification steps are:

1. **Image Acquisition:** The first step in image acquisition is to extract the image from the field by capturing it from the camera and dividing it into the media folder into 10 classes of different types of leaf image diseases.
2. **Image Pre-Processing:** It is the second step in image pre-processing that helps you enhance the image color from RGB to Gray and also includes conventional techniques like image dimension resizing and filtering the noise.
3. **Image Segmentation:** Image segmentation is the third step that helps you to segment images into different parts and create multiple segmented images, like creating a digital image with multiple segments.

4. **Feature Extraction:** Step four is featuring extraction that helps to extract multiple features from a given image like if we have a single input image feature extraction that helps to create multiple frames of a given image.
5. **Classification:** Classification is used for classifying your data into train and test classes, and with the help of fully connected layers, we can easily classify given data with multiple classes of leaf disease.

6.4 COMPARATIVE STUDY OF DIFFERENT TOMATO LEAF DISEASE AND CLASSIFICATION TECHNIQUES USING CNN (TABLES 6.1 AND 6.2)

TABLE 6.1 Comparison of Tomato Leaf Prediction and Classification Techniques-I

Author	Title	Proposed Technique	Dataset
Agarwal M. Singh	ToLed	CNN+ supervise learning.	Plant leaves dataset.

TABLE 6.2 Comparison of Tomato Leaf Prediction and Classification Techniques-II

Author	Accuracy	Benefits	Limitations
Agarwal M. Singh	91.2%	It combines clustering.	Tomato disease type can't be classified.

6.5 CONCLUSION

The performance of CNNs in object acknowledgment and picture grouping has gained tremendous headway in the past couple of years. Already, the conventional methodology for picture arrangement tasks has been founded alongside designed components, like SIFT, HoG, SURF, and so on, and afterward to utilize some type of learning calculation in these element spaces. The presentation of these methodologies in this way relied intensely upon the basic predefined highlights. Component engineering itself is a mind boggling and dreary interaction which should be returned to each time the current issue or the related dataset changes significantly. This issue happens in all conventional endeavors to recognize plant diseases utilizing PC vision as they incline intensely close by designed components, picture improvement strategies, and a large group of other complex and work concentrated systems. There many created techniques in the detection and arrangement of plant diseases utilizing infected leaves of plants. Nonetheless, there is still no efficient and effective business solution that can be utilized to recognize the diseases in our works, we used CNN approach, i.e., CNN model for the make out of plant diseases using healthy and diseased tomato leaf images. To train and test the model, we collected the data from a new plant disease dataset image dataset containing healthy and unhealthy tomato crop leaves. The dataset contains 21,416 images, from which 18,345 images are for training the model and 3,071 are testing images. After splitting the dataset into 80–20 (80% of whole data for training, 20% whole images for testing), we obtained the best accuracy rate of 96.69% in the CNN model. On average, short time was required to train the images in the CNN model and it took 10 epochs, respectively, on colored images. In comparison with other deep-learning outcomes, the implemented deep-learning model has better prognostic capability in terms of both accuracy and loss.

KEYWORDS

- **artificial intelligence**
- **convolutional neural network**
- **crop disease**
- **deep learning**
- **Septoria**
- **tomato leaf**
- **tomato leaf disease**

REFERENCES

1. Agarwal, M., Singh, A., Arjaria, S., Sinha, A., & Gupta, S., (2020). ToLeD: Tomato leaf disease detection using convolutional neural network. *Procedia Computer Science, 167*, 293–301.
2. Sannakki, S. S., Vijay, S. R., Nargund, V. B., Parag, K., Sanjeev, S. S., Vijay, S. R., & Vijay, S., (2013). Using neural networks to diagnose and classify grape leaf diseases. In: *2013 Fourth International Conference on Computing, Communications, and Networking Technologies (ICCCNT)* (pp. 1–5). IEEE.

3. Aida, S., Suhaili, B., Noor, E. A., Habibah, H., & Kutty, (2013). Using neural network analysis to classify watermelon leaf diseases. In: *IEEE Business Engineering and Industrial Applications Colloquium (BELAC), 2013* (pp. 459–464). 2013 IEEE.
4. Arjun, K. M., (2013). Indian agriculture status, importance, and role in the Indian economy. *International Journal of Agriculture and Food Science Technology*, *4*(4), 343–346.
5. Shrivastava, S., & Hooda, D. S., (2014). Automatic brown spot and frog eye detection from the image captured in the field. *American Journal of Intelligent Systems*, *4*(4), 131–134.
6. Zhang, M., & Meng, Q., (2011). Automatic citrus canker detection from leaf images captured in the field. *Pattern Recognition Letters, 32*(15), 2036–2046.
7. Jadhav, S. B., (2019). Convolutional neural networks for leaf image-based plant disease classification. *IAES International Journal of Artificial Intelligence*, *8*(4), 328.
8. Dash, J., Verma, S., Dasmunshi, S., & Nigam, S., (2018). Plant health monitoring system using raspberry Pi. *International Journal of Pure and Applied Mathematics*, *119*(15), 955–959.
9. Patil, B., Panchal, H., Yadav, M. S., Singh, M. A., & Patil, M. D., (2017). Plant monitoring using image processing, raspberry Pi & IoT. *International Research Journal of Engineering and Technology (IRJET)*, *4*(10).
10. Mhatre, R., & Lanke, V. (2021). *Cotton Leaves Disease Detection and Cure Using Deep Learning.*
11. Sachin, D. K., & Patil, A. B., (2015). Plant disease detection using image processing. *ICCUBEA (International Conference on Computing, Communication, Control, and Automation) is a Biennial International Conference on Computing, Communication, Control, and Automation.)* (pp. 768–771). IEEE.
12. Brahimi, M., Arsenovic, M., Laraba, S., Sladojevic, S., Boukhalfa, K., & Moussaoui, A., (2018). Deep learning for plant diseases: Detection and saliency map visualization. In *Human and Machine Learning* (pp. 93–117). Springer, Cham.

CHAPTER 7

ANALYSIS OF M[X]/G/1 RETRIAL QUEUING MODEL WITH BERNOULLI WORKING VACATION, VACATION INTERRUPTION, AND ORBITAL SEARCH WITH FUZZY PARAMETERS

J. EBENESAR ANNA BAGYAM,[1] R. S. DHARANI,[1] MANOJ KUMAR CHANDE,[2] A. VISWANATHAN,[3] C. BEULAH CHRISTALIN LATHA,[4] and B. VENNILA[1]

[1]*Department of Mathematics, Avinashilingam Institute for Home Science and Higher Education for Women, Coimbatore, Tamil Nadu, India*

[2]*Department of Applied Mathematics, Shri Shankaracharya Institute of Professional Management and Technology, Raipur, Chhattisgarh, India*

[3]*Department of Mathematics, SNS College of Technology, Coimbatore, Tamil Nadu, India*

[4]*Department of Digital Sciences, Karunya Institute of Technology and Sciences, Coimbatore, Tamil Nadu, India*

ABSTRACT

The research analyzes an M[x]/G/1 retrial queueing model that incorporates Bernoulli working vacation, vacation interruption, and orbital search with fuzzy parameters. Fuzzy logic is employed due to its ability to handle partial truths and uncertainties inherent in real-world scenarios. The study explores

Computational Optimization, Modeling, and Simulation for Engineering Applications.
Anupam Shukla, Sourabh Rungta, Mohan Awasthy, and Rakesh Himte (Eds.)

various aspects, including system assumptions, steady-state governing equations, and performance measures. It introduces notations and equations governing the model, emphasizing the utilization of fuzzy numbers to represent uncertain parameters. Performance measures such as server idle probability, busy probability, and vacation probability are derived and analyzed. The study also investigates the mean number of customers in the orbit and presents membership functions for fuzzy constraints. Through extensive analysis and modeling, the research contributes to understanding and optimizing queueing systems under fuzzy environments.

7.1 INTRODUCTION

Fuzzy logic is a kind of mathematical logic in which truth can take on a range of values ranging from 0 to 1. Fuzzy logic enables a statement to be both partially true and partially incorrect at the same time. Real-world circumstances are not sharp and predictable and thus cannot be adequately articulated. The requirement for a conceptual framework that can address the issue of uncertainty and lexical imprecision drove the creation of fuzzy logic in significant part. Fuzzy set theory is a valuable tool for simulating the type of uncertainty that comes with ambiguity, imprecision, and a lack of understanding about a specific aspect of the situation at hand.

Zadeh [1] was the first to develop the notion of fuzzy sets in his paper titled "Fuzzy sets." Zadeh's extension concept is a milestone in fuzzy set theory. Many academics have explored the topic of fuzzy queuing systems after Zadeh.

Many researchers are showing their interest in analyzing queuing models using fuzzy attributes in recent years. Ke et al. [2] created a system characteristics membership function with fuzzy parameters in a retrial queuing model. Barak & Fallahnezhad [3] performed sensitivity analysis to compare fuzzy queuing models of single server Markovian queue and Erlangian queue. Noora et al. [4] studied fuzzy multiserver retrial queue with finite source. Gomathi & Adaikalam [5] developed the model to improve the efficiency of the OFQS switch design. Visalakshi & Savitha [6] discussed the significant role of the pentagonal fuzzy number concept in the FM/FM/1 queuing model. Bagyam & Udayachandrika [7] studied a fuzzy admission control bulk arrival vacation two-phase retrial queuing model. Kannadasan & Padmavathi [8] utilized hexagonal fuzzy numbers to investigate the traditional fuzzy retry queue with a working vacation model. Bouchentouf et al.

[9] used the stationary distribution to analyze a limited capacity single server feedback queuing system, including vacation and non-persistent customers, and they found crucial metrics of the model's efficacy. To deal with uncertainty, Nayeri et al. [10] used queuing theory and adaptable fuzzy stochastic optimization. Mishra et al. [11] investigated the fuzzy cost optimization in control queuing models with a single server.

7.2 ASSUMPTIONS AND DESCRIPTION OF THE MODEL

Consider a single server bulk arrival retrial queueing system with server working vacation. If the server is accessible when a new batch arrives, one of the customers starts his service instantly, and the remaining customers enter the orbit. If a new or retry customer comes at the server's idle state, it immediately restarts in normal mode. When the orbit gets empty, the server goes on a working vacation. The working vacation time is a slower-paced operational phase with parameter α. If a customer arrives during a vacation, the server begins to perform at a decreased service rate, referred to as vacation mode. If there are clients in orbit at a service completion instant during the working vacation time, the server will end the vacation and resume normal busy operations. If there are no customers in the system at the completion of the vacation, the server either rejoins the system and stays idle to service a new customer in normal mode with probability p (single working vacation) or departs for another vacation with the complementary probability (multiple working vacations). If the orbit is not empty after vacation, the server looks for customers in orbit with probability θ or remains idle with complementary probability.

Define the following notations:

- **λ:** Arrival rate, C (ς) , m_1 and m_2– PGF of batch size and first two moments.
- **$R_T*(\theta)$ and $\omega(\kappa)$:**LST and hazard rate of inter retrial time.
- **$\mu_b(\kappa)$, $S_b*(\theta)$, γ_1 and γ_2:** Hazard rate, LST, and first two moments of normal service rate.
- **$\mu_v(\kappa)$, $S_v*(\theta)$ and β_1 and β_2:** Hazard rate, LST, and first two moments of working vacation rate.

Define the states of the server at time τ as, $C(\tau) = 0, 1, 2$, and 3 if the server is idle on regular mode (I_n), idle on vacation mode (Iv_0), busy on regular mode (P_n) and busy on vacation mode (V_n), respectively, where n denotes number of customers in the orbit. Let I_0, $I_{v0,}$ $I_n(\kappa)$, $P_n(\kappa)$, $V_n(\kappa)$ are the corresponding steady state probabilities.

7.3 STEADY STATE GOVERNING EQUATION

The system of steady state equations that governs the model under SVT is given below:

$$\lambda I_0 = \alpha\, p\, I_{v0} \tag{1}$$

$$(\lambda + a)\, I_{v0} = \alpha q\, I_{v0} + \int_0^\infty P_0\,(\kappa)\, \mu_b(\kappa)\, d\kappa + \int_0^\infty V_{0\,(\kappa)}\, \mu_v(\kappa)\, d\kappa \tag{2}$$

$$\frac{d}{d\kappa} I_n(\kappa) = -\,(\lambda + \varpi(\kappa))\, I_n(\kappa),\ n \geq 1 \tag{3}$$

$$\frac{d}{d\kappa} P_0(\kappa) = -\,(\lambda + \mu_b(\kappa))\, P_0(\kappa) \tag{4}$$

$$\frac{d}{d\kappa} P_n(\kappa) = -\,(\lambda + \mu_b(\kappa))\, P_n(\kappa) + \lambda \sum_{\kappa=1}^{n} c_\kappa\ P_{n-k}(k),\ n \geq 1 \tag{5}$$

$$\frac{d}{d\kappa} V_0(\kappa) = -\,(\lambda + \alpha + \mu_v(\kappa))\, V_0(\kappa) \tag{6}$$

$$\frac{d}{d\kappa} V_n(\kappa) = -\,(\lambda + \alpha + \mu_v(\kappa))\, V_n(\kappa) + \lambda \sum_{\kappa=1}^{n} c_\kappa\, V_{n-k}(k)\ n \geq 1 \tag{7}$$

The boundary conditions are:

$$I_n(0) = (1-\theta) \int_0^\infty P_n\,(\kappa)\ \mu_b(\kappa)\, d\kappa + \int_0^\infty v_n\,(\kappa)\, \mu_v(\kappa)\, d\kappa,\ n \geq 1 \tag{8}$$

$$P_0(0) = \int_0^\infty I_1\,(\kappa)\, \varpi(\kappa) d\kappa + \theta \int_0^\infty P_1\,(\kappa)\, \mu_b(\kappa)\, d\kappa + \alpha \int_0^\infty v_0\,(\kappa)\ d\kappa + \lambda C_1 I_0 \tag{9}$$

$$P_n(0) = \lambda C_{n+1} I0 \int_0^\infty I_{n+1}\,(\kappa)\ \varpi(\kappa) d\kappa + \lambda \sum_{\kappa=1}^{n} C_\kappa\ \int_0^\infty I_{n-k+1}\,(\kappa)\, \mu\kappa + \alpha \int_0^\infty P_{n+1}\,(\kappa)$$
$$\mu_b(\kappa) d\kappa + \alpha \int_0^\infty v_n\,(\kappa)\ d\kappa,\ n \geq 1 \tag{10}$$

$$V_n(0) = \lambda I_{v0} \tag{11}$$

Define P.G.F. for $|\varsigma| \leq 1$ as $I(\kappa, \varsigma) = \sum_{n=1}^{\infty} I_n(\kappa)\varsigma^n$; $P(\kappa, \varsigma) = \sum_{n=0}^{\infty} p_n(\kappa)\varsigma^n$; $V(\kappa, \varsigma) = \sum_{n=1}^{\infty} V_n(\kappa)\varsigma^n$;

Using the PGF method, we get the following equations:

$$I(\kappa, \varsigma) = e^{-\lambda\kappa}(1 - R_T(\kappa)) \{\varsigma\ I_{v0} [((1-\theta)\ S_b^*(A_b(\varsigma))(\lambda g(\varsigma) + C(\varsigma)\alpha p) + (\lambda(S_v^*(A_v(\varsigma) - 1) - \alpha p))] - \theta^{I_{v0}} [((\lambda(S_v^*(A_v(\varsigma) - 1) - \alpha p))\ S_b^*(A_b(\varsigma))]\}/Dr \tag{12}$$

$$P(\kappa, \varsigma) = e^{-A_b(\varsigma)k} (1 - S_b(A_b(\varsigma)))\ [\varsigma C(\varsigma)\alpha p\, I_{v0} + \lambda I_{v0} (S_v^*(A_v(\varsigma)) (R_T^*(\lambda) + \varsigma C(\varsigma)(1 - R_T^*(\lambda)) + \varsigma g(\varsigma)] - (\lambda + \alpha p)\ I_{v0} (R_T^*(\lambda) + \varsigma C(\varsigma) (1 - R_T^*(\lambda)))]/Dr \tag{13}$$

$$V(\kappa, \varsigma) = \lambda I_{v0}\ e^{-A_b(\varsigma)k} (1 - S_v(A_v(\varsigma))) \tag{14}$$

The partial P.G.F. of the orbit size when the server is idle is:

$$I(\varsigma) = \left(\frac{1 - R_T^*(\lambda)}{\lambda}\right) \{\varsigma\ I_{v0}\ [((1-\theta)\ S_b^*(A_b(\varsigma))(\lambda g(\varsigma) + C(\varsigma)\ \alpha p) + (\lambda(S_v^*(A_v(\varsigma)) - 1) - \alpha p))] - \theta^{I_{v0}} [(\lambda(S_v^*(A_v(\varsigma)) - 1) - \alpha p))\ S_b^*(A_b(\varsigma))]\}/Dr \tag{15}$$

The partial P.G.F. of the orbit size when the server is busy is:

$$P(\varsigma) = \left(\frac{1 - S_b^*(A_b(\varsigma))}{A_b(\varsigma)}\right) [\varsigma C(\varsigma)\ \alpha p\ I_{v0} + \lambda\ I_{v0}\ (S_v^*(A_v(\varsigma))\ (R_T^*(\lambda) + \varsigma c(\varsigma)(1 - R_T^*(\lambda)) + \varsigma\ g(\varsigma) - (\lambda + \alpha p)\ (R_T^*(\lambda) + \varsigma C(\varsigma)(1 - R_T^*(\lambda)))]\ /Dr \tag{16}$$

The partial P.G.F. of the orbit size when the server is on vacation is:

$$V(\varsigma) = \frac{\lambda I v_0}{\alpha} g(\varsigma) \tag{17}$$

The partial P.G.F. of the orbit size is:

$$P_q(\varsigma) = I_{v0} \{(1 - C(\varsigma))[(\frac{\lambda}{\alpha}(\alpha + \lambda g(\varsigma)) + \alpha p)\ (\varsigma - ((1-\theta)\ (R_T^*(\lambda) + \varsigma C(\varsigma)(1 - R_T^*(\lambda)) + \theta)\ S_b^*(A_b(\varsigma))) + \varsigma(1 - R_T^*(\lambda))\ [(1-\theta)(\lambda g(\varsigma) + C(\varsigma)\alpha p)\ S_b^*(A_b(\varsigma)) + (\lambda(S_v^*(A_v(\varsigma) - 1) - \alpha p)] - \theta(1 - R_T^*(\lambda))\ (\lambda(S_v^*(A_v(\varsigma) - 1) - \alpha p))\ S_b^*(A_b(\varsigma)] + (1 - S_b^*(A_b(\varsigma)))[(\lambda(S_v^*(A_v(\varsigma) - 1) - \alpha p))\ (R_T^*(\lambda) + \varsigma C(\varsigma)(1 - R_T^*(\lambda))) + \varsigma(C(\varsigma)\alpha p + \lambda g(\varsigma))]\}/Dr_1 \tag{18}$$

where;

$$Dr_1 = A_b(\varsigma)\,(\varsigma-((1-\theta)\,(R_T^*(\lambda) + \varsigma C(\varsigma)\,(1 - R_T^*(\lambda))) + \theta)\,S_b^*(A_b(\varsigma)))$$

7.4 PERFORMANCE MEASURES

The probability that the server is idle is:

$$I = I_{v0}\,(1 - R_T^*(\lambda))\,[m_1((\lambda(1 - S_v^*(a))+_\gamma \alpha p)_{\gamma 1}) - \theta(1 - S_v^*(a) + \frac{\alpha p}{\lambda}(1-\theta)\,(\frac{\lambda}{\alpha} m_1(1 - S_v^*(a))+\frac{m_1 ap}{\lambda})]\,/(1-\theta)\,R_T^*(\lambda)) + \theta - (1-\theta)\,m_1(1 - R_T^*(\lambda)) - \lambda m_{1\gamma 1} \quad (19)$$

The probability that the server is busy is:

$$P = I_{v0\,\gamma 1}[(\lambda(1 - S_v^*(a) + \alpha\,R_T^*(\lambda) + \lambda\,m_1\,(1 - S_v^*(a)(\frac{\lambda}{\alpha} - 1) + m_1(\lambda(1 - S_v^*(a) + \alpha p)\,R_T^*(\lambda)\,]/(1-\theta)\,R_T^*(\lambda) + \theta - (1-\theta)\,m_1\,(1 - R_T^*(\lambda)) - \lambda\,m_{1\gamma 1} \quad (20)$$

The Probability that the server is on vacation is:

$$V = \frac{\lambda v_0}{a}(1 - S_v^*(a) \quad (21)$$

Using the normalizing condition, we get:

$$I_0 = \alpha p[(1-\theta)\,R_T^*(\lambda) + \theta - (1-\theta)\,m_1\,(1 - R_T^*(\lambda)) - \lambda m_{1\gamma 1}]/\lambda[(1-\theta)\,\frac{\lambda}{\alpha}(1 - S_v^*(a))\,R_T^*(\lambda) + (\frac{ap}{\lambda}+1)\,R_T^*(\lambda) - (1-\theta)\,m_1\,(1 - R_T^*(\lambda)) - \lambda m_{1\gamma 1} + \theta(1 - R_T^*(\lambda))\,S_v^*(a) + \alpha p_{\gamma 1}\,R_T^*(\lambda) + \frac{\lambda\theta}{a}(1 - S_v^*(a))] \quad (22)$$

$$I_{v0} = [(1-\theta)\,R_T^*(\lambda) + \theta - (1-\theta)\,m_1\,(1 - R_T^*(\lambda)) - \lambda m_{1\gamma 1}]/[(1-\theta)\,\frac{\lambda}{\alpha}(1 - S_v^*(a))\,R_T^*(\lambda) + (\frac{ap}{\lambda}+1)\,R_T^*(\lambda)) - (1-\theta)\,m_1\,(1 - R_T^*(\lambda)) - \lambda m_{1\gamma 1} + \theta(1 - R_T^*(\lambda))\,S_v^*(a) + \alpha p_{\gamma 1}\,R_T^*(\lambda) + \frac{\lambda\theta}{a}(1 - S_v^*(a))] \quad (23)$$

The mean number of customers in the orbit is:

$$L_q = \lim_{\varsigma \to 1}\frac{d}{d\varsigma}p_q(\varsigma) = I_{v0}\,[Nr_1'''(1)\,Dr_1''(1) - Dr_1'''(1)\,Nr_1''(1)/3(Dr_1''(1))^2$$

where;

$Nr_1''(1) = I_{v0}\{-2m_1[\lambda + \frac{\lambda^2}{\alpha}(1 - S_v^*(\alpha) + \alpha p - \lambda^2 m_1\gamma_1 - (1-\theta)\lambda(m_1+1)(1 - R_T^*(\lambda)) - (1-\theta)\frac{\lambda^2}{\alpha}(1 - S_v^*(\alpha))(1 - R_T^*(\lambda)) - (1-\theta)\alpha p(1 - R_T^*(\lambda)) + \lambda^2\gamma_1 R_T^*(\lambda)(1 - S_v^*(\alpha)) + \lambda\gamma_1\alpha p]\}$

$Nr_1'''(1) = I_{v0}\{-3\lambda m_1[\frac{2\lambda}{\alpha} r_1 - \frac{2\lambda^2}{\alpha} m_1\gamma_1 r_1 - (1-\theta)\frac{2\lambda}{\alpha} r_1 T_1 - 2(1-\theta)\lambda m_1\gamma_1 T_1 - \frac{\lambda^3}{\alpha} m_2\gamma_2(1 - S_v^*(\alpha)) - 2(1-\theta)\frac{\lambda^2}{\alpha} m_1\gamma_1 T_1(1 - S_v^*(\alpha)) - (1-\theta)\frac{\lambda}{\alpha} T_2(1 - S_v^*(\alpha)) - \lambda m_2\gamma_2\alpha p - 2(1-\theta)m_1\gamma_1 T_1\alpha p - (1-\theta)T_2(\frac{\alpha p}{\lambda} + 1) + 2(1-\theta)\tau_1(1 - R_T^*(\lambda)) + 2(1-\theta)\frac{m_1\alpha p}{\lambda}(1 - R_T^*(\lambda)) + 2\lambda m_1\gamma_1(1 - S_v^*(\alpha))(1 - R_T^*(\lambda)) + 2\alpha p m_1\gamma_1(1 - R_T^*(\lambda)) + 2\lambda m_1\beta_1(1 - R_T^*(\lambda)) - 2\theta\lambda m_1\beta_1(1 - R_T^*(\lambda)) + (1-\theta)\tau_2(1 - R_T^*(\lambda)) + (1-\theta)\frac{m_2\alpha p}{\lambda}(1 - R_T^*(\lambda)) + 2(1-\theta)\lambda\tau_1 m_1\gamma_1(1 - R_T^*(\lambda)) + 2(1-\theta)m_1^2\gamma_1\alpha p(1 - R_T^*(\lambda)) - \lambda^2 m_2\gamma_2 S_v^*(\alpha)(1 - R_T^*(\lambda)) - \gamma^2 m_2\gamma_2 R_T^*(\lambda) + (1-\theta)\lambda m_2\gamma_2(1 - R_T^*(\lambda)) - 2\theta\lambda^2 m_1^2\gamma_1\beta_1(1 - R_T^*(\lambda)) + \theta\alpha p\lambda m_2\gamma_2(1 - R_T^*(\lambda)) - 2\theta\lambda^2 m_1^2\gamma_1\beta_1(1 - R_T^*(\lambda)) + \theta\alpha p\lambda m_2\gamma_2(1 - R_T^*(\lambda)) + \lambda^3\gamma_1 m_2\beta_2 + 2\lambda^2\gamma_1\beta_1 T_1 - \lambda\gamma_1 T_2(1 - S_v^*(\alpha)) - \alpha p\gamma_1 T_2 + m_2\alpha p\gamma_1 + \lambda\gamma_1\tau_2 + 2m_1\alpha p\gamma_1 + 2\lambda\gamma_1\tau_1] - 3m_2[\lambda - (1-\theta)\lambda T_1\frac{\lambda^2}{\alpha}(1 - S_v^*(\alpha)) - \frac{\lambda^3}{\alpha} m_1\gamma_1(1 - S_v^*(\alpha)) + \alpha p - (1-\theta)\lambda p T_1 - (1-\theta)\frac{\lambda^2}{\alpha}(1 - S_v^*(\alpha))T_1 - \lambda^2 m_1\gamma_1 S_v^*(\alpha))(1 R_T^*(\lambda) - \lambda m_1\gamma_1\alpha p R_T^*(\lambda) + (1-\theta)\lambda T_1(1 - R_T^*(\lambda)) + (1-\theta)m_1\alpha p(1 - R_T^*(\lambda)) + \lambda^4 m_1\beta_1\gamma_2 + \lambda^2 m_1\beta_1(1 - R_T^*(\lambda)) + \lambda^2\gamma_2\alpha p(1 - T_1) + \lambda^2\gamma_2 m_1\alpha p + \lambda^2\gamma_2(1 - S_v^*(\alpha))(1 - T_1) + \lambda^3\gamma_2\tau_1]\}$

$Dr_1'''(1) = -2\lambda m_1[1 - \lambda\gamma_1 m_1 + (1-\theta)(m_1+1)(1 - R_T^*(\lambda))]$

$Dr_1''(1) = -3\lambda\{-m_1[\lambda^2 m_2\gamma_2 + 2(1-\theta)\lambda m_1\gamma_1 T_1 + (1-\theta)T_2] + m_2[1 - \lambda\gamma_1 m_1 - (1-\theta)T_1]\}$ $\tau_1 = g'(1) = \frac{\lambda}{\alpha} m_1(1 - S_v^*(\alpha)) - \lambda m_1\beta_1$ $\tau_2 = g''(1) = \frac{2\lambda^2}{a^2} m_1^2(1 - S_v^*(\alpha)) - \frac{2\lambda^2}{a^2} m_1^2\beta_1 + \frac{\lambda}{\alpha} m_2(1 - S_v^*(\alpha)) - \lambda m_2\beta_2$ $T_1 = (1 - R_T^*(\lambda)) + m_1(1 - R_T^*(\lambda))$ and $T_2 = 2m_1(1 - R_T^*(\lambda)) + m_2(1 - R_T^*(\lambda))$

7.5 ANALYSIS OF MODEL WITH FUZZY PARAMETERS

The membership function (M.F.) of the mean number of customers in the orbit for the model under study with fuzzy constraints is presented in this section.

Assume that retrial and service time (both normal and vacation mode) all follow the exponential distribution with parameters ω, μ_b and μ_v, respectively.

Suppose the parameters λ, ω, μ_b and μ_v is approximately identified and are denoted by fuzzy numbers $\widehat{\lambda}\widehat{\omega}$, $\widehat{\mu}_b$ and $\widehat{\mu}_v$ respectively. Let $\psi_{\widehat{\lambda}}(b)$, $\psi_{\widehat{\omega}}(d)$, $\widehat{\mu}_b(e)$ and $\psi\widehat{\mu}_v(f)$ and represent the M.F.'s of the equivalent parameters. Then, $\widehat{\lambda}=\{(b, \psi_{\widehat{\lambda}}(b))\,/\,b \in B\}, \widehat{\omega}=\{(d, \psi_{\widehat{\omega}}(d))/d \in D\}, \widehat{\mu}_b=\{(e.\ \psi_{\widehat{\mu}_b}(e))/e \in E\}$ and $\widehat{\mu}_v=\{(f, \psi_{\widehat{\mu}v}(f)/f \in F\}$, where D, B, E, and F are the crisp total set of retrial, arrival, service rates in regular and vacation mode, respectively.

As $\widehat{\lambda}$, $\widehat{\omega}$, $\widehat{\mu}_b$ and $\widehat{\mu}_v$ are all fuzzy numbers $L_q(\widehat{\lambda}, \widehat{\omega}, \widehat{\mu}_b, \widehat{\mu}_v)$ is also a fuzzy number. Based on the concept of extension principle, the M.F. $L_q(\widehat{\lambda}, \widehat{\omega}, \widehat{\mu}_b, \widehat{\mu}_v)$ is defined as:

$$\psi_{L_q}(\widehat{\lambda}, \widehat{\omega}, \widehat{\mu}_b, \widehat{\mu}_v)(\varsigma)=\psi_{\widehat{L}_q}(\varsigma)=\sup_{\Omega} \min\{\psi_{\widehat{\lambda}}(b), \psi_{\widehat{\omega}}(d), \psi_{\mu_b}(e), \psi_{\mu_v}(f)\ /z=L_q(b, d, e, f)\} \tag{24}$$

where; $\Omega=\{b \in B, d \in D, e \in E, f \in F\}$

$\alpha p[(1-\theta) R_T^*(b)+\theta-(1-\theta) m_1(1-R_T^*-bm_1\gamma_1]/b[(1-\theta)\frac{b}{a}(1-S_V^*(\alpha)) R_T^*(b)+(\frac{\alpha p}{b}+1) R_T^*(b)-(1-\theta) m_1(1-R_T^*(b))-bm_1\gamma_1+\theta(1-R_T^*(b)) S_V^*(\alpha)+\alpha p\gamma_1 R_T^*(b)+\frac{b\theta}{\alpha}(1-S_V^*(\alpha))]>0$ and $[(1-\theta) R_T^*(b)+\theta-(1-\theta) m_1(1-R_T^*(b))-bm_1\gamma_1]/[(1-\theta)\frac{b}{a}(1-S_V^*(\alpha) R_T^*(b)+(\frac{\alpha p}{b}+1) R_T^*(b)-(1-\theta) m_1(1-R_T^*(b))-bm_1\gamma_1+\theta(1-S_V^*(\alpha) S_V^*(\alpha)+\alpha p\gamma_1 R_T^*(b)+\frac{b\theta}{\alpha}(1-S_V^*(\alpha))]>0\}$ and

$$L_q=I_{v0}[Nr_3(1)\, Dr_2(1)-Dr_3(1)\, Nr_2(1)]/3(Dr_2(1)) \tag{25}$$

$Nr_2(1)=I_{v0}\{-2m_1[b+\frac{b^2}{a}(1-S_V^*(\alpha)+\alpha p-b^2m_1\gamma_1-(1-\theta)\, b(m_1+1)(1-R_T^*(b))-(1-\theta)\frac{b^2}{a}(1-S_v^*(a))(1-R_T^*(b))-(1-\theta)\, \alpha p(1-R_T^*(b))+b^2\gamma_1 R_T^*(b)(1-S_V^*(\alpha))+b\gamma_1\alpha p]\}$

$Nr_3(1)=I_{v0}\{-3bm_1[\frac{2b}{a}\tau_1-\frac{2b^2}{a}m_1\gamma_1\tau_1-(1-\theta)\frac{2b}{a}\tau_1 T_1-2(1-\theta)\, bm_1\gamma_1 T_1-\frac{b^3}{a}m_2\gamma_2(1-S_v^*(\alpha))-2(1-\theta)\frac{b^2}{a}m_1\gamma_1 T_1(1-S_v^*(\alpha))-(1-\theta)\, S_v^* T_2(1-S_v^*(\alpha))-bm_2\gamma_2\alpha p-2(1-\theta)$

$m_1\gamma_1T_1\alpha p - (1-\theta)\,T_2(\frac{\alpha p}{b}+1) + 2(1-\theta)\,\tau_1(1-R_T^*(b)) + 2(1-\theta)\,\frac{m_1\alpha p}{b}(1-R_T^*(b)) + 2bm_1\gamma_1\,(1-S_V^*(\alpha))\,(1-R_T^*(b)) + 2\alpha pm_1\gamma_1(1-R_T^*(b)) + 2bm_1\beta_1(1-R_T^*(b)) - 2\theta bm_1\beta_1(1-R_T^*(b)) + (1-\theta)\,\tau_2\,(1-R_T^*(b)) + (1-\theta)\,\frac{m_2\alpha p}{b}(1-R_T^*(b)) + 2(1-\theta)\,b\tau_1m_1\gamma_1(1-R_T^*(b)) + 2(1-\theta)\,m_1^2\gamma_1\alpha p\,(1-R_T^*(b)) - b^2m_2\gamma_2R_T^*(b))\,(1-R_T^*(b)) - b^2m_2\gamma_2R_T^*(b) + (1-\theta)\,bm_2\gamma_2\,(1-R_T^*(b)) - 2\theta b^2m_1^2\gamma_1\beta_1\,(1-R_T^*(b)) + \theta\alpha pbm_2\gamma_2\,(1-R_T^*(b)) + b^3\gamma_1m_2\beta_2 + 2b^2\gamma_1\beta_1T_1 - b\gamma_1T_2\,(1-S_V^*(\alpha)) - \alpha p\gamma_1T_2 + m_2\alpha p\gamma_1 + b\gamma_1\tau_2 + 2m_1\alpha p\gamma_1 + 2b\gamma_1\tau_1] - 3m_2[b-(1-\theta)\,b\,T_1 + \frac{b^2}{\alpha}(1-S_V^*(\alpha)) - \frac{b^3}{\alpha}\,m_1\gamma_1(1-S_V^*(\alpha)) + \alpha p - (1-\theta)\,\alpha pT_1 - (1-\theta)\,\frac{b^2}{\alpha}(1-S_V^*(\alpha))\,T_1 - b^2m_1\gamma_1S_V^*(\alpha)(1-R_T^*(b)) - bm_1\gamma_1\alpha p\,R_T^*(b) + (1-\theta)\,b\tau_1(1-R_T^*(b)) + (1-\theta)\,m_1\alpha p(1-R_T^*(b)) + b^4m_1\beta_1\gamma_2 + b^2m_1\beta_1(1-R_T^*(b)) + b^2\gamma_2\alpha p(1-T_1) + b^2\gamma_2m_1\alpha p + b^2\gamma_2\,(1-S_V^*(\alpha))\,(1-T_1) + b^3\gamma_2\tau_1]\}\ Dr_2(1) = -2bm_1[1-b\gamma_1m_1 + (1-\theta)\,(m_1+1)\,(1-R_T^*(b))]\ Dr_3(1) = -3b\{-m_1[b^2m_2\gamma_2 + 2(1-\theta)\,bm_1\gamma_1T_1 + (1-\theta)\,T_2] + m_2[1-b\gamma_1m_1-(1-\theta)\,T_1]\}\tau_1 = \frac{b}{a}\,m_1(1-S_V^*(\alpha)) - bm_1\beta_1$, $\tau_2 = \frac{2b^2}{\alpha^2}\,m_1^2(1-S_V^*(\alpha)) - \frac{2b^2}{\alpha^2}\,m_1^2\beta_1 + \frac{b}{a}\,m_2(1-S_V^*(\alpha)) - bm_2\beta_2\ T_1 = (1-R_T^*(b)) + m_1(1-R_T^*(b))$ and $T_2 = 2m_1(1-R_T^*(b)) + m_2(1-R_T^*(b))$

The α-cuts of $\hat{\lambda}$, $\hat{\omega}$, $\hat{\mu}_b$, $\hat{\mu}_v$ are crisp intervals defined as α:

$$\lambda(\alpha) = [\min_{b\in B}\{b/\Psi_{\hat{\lambda}}(b)\geq\alpha\},\ \max_{b\in B}\{b/\Psi_{\hat{\lambda}}(b)\geq\alpha\}] = b_\alpha^L, b_\alpha^U \tag{26}$$

$$\omega(\alpha) = [\min_{d\in D}\{d/\Psi_{\hat{\omega}}(d)\geq\alpha\},\ \max_{d\in D}\{d/\Psi_{\hat{\omega}}(d)\geq\alpha\}] = d_\alpha^L, d_\alpha^U \tag{27}$$

$$\mu_b(\alpha) = [\min_{e\in E}\{e/\psi_{\mu_b}(e)\geq\alpha\},\ \max_{e\in E}\{e/\psi_{\mu_b}(e)\geq\alpha\}] = e_\alpha^L, e_\alpha^U \tag{28}$$

$$\mu_v(\alpha) = [\min_{f\in F}\{f/\psi_{\mu_v}(f)\geq\alpha\},\ \max_{f\in F}\{f/\psi_{\mu_v}(f)\geq\alpha\}] = f_\alpha^L, f_\alpha^U \tag{29}$$

From Eqns. (26)–(29) are the functions of alpha and their lower and upper limits are displayed below:

$$b_\alpha^L = \text{Min}\ \psi_{\hat{\lambda}}^{-1}(\alpha),\quad b_\alpha^U = \text{Max}\ \psi_{\hat{\lambda}}^{-1}(\alpha),\quad d_\alpha^L = \text{Min}\ \psi_{\hat{\omega}}^{-1}(\alpha),\quad D_\alpha^U = \text{Max}\ \psi_{\hat{\omega}}^{-1}(\alpha)$$

$$e_\alpha^L = \text{Min}\ \psi_{\hat{\mu}_b}^{-1}(\alpha),\quad e_\alpha^U = \text{Max}\ \psi_{\hat{\mu}_b}^{-1}(\alpha),\quad f_\alpha^L = \text{Min}\ \psi_{\hat{\mu}_v}^{-1}(\alpha),\quad f_\alpha^U = \text{Max}\ \psi_{\hat{\mu}_v}^{-1}(\alpha)$$

The alpha cut approach is used to construct the M.F. $\psi_{L_q}(\varsigma)$, which is dependent on α. According to Zadeh's extension concept, $\psi_{\hat{L}_q}(\varsigma)$ is the minimum of $\psi_{\hat{\lambda}}(b), \psi_{\hat{\omega}}(d), \psi_{\hat{\mu}_b}(e)$ and $\psi_{\hat{\mu}_v}(f)$. To obtain $\psi_{\hat{L}_q}(\varsigma)$, Let us assume one of $\psi_{\hat{\lambda}}(b), \psi_{\hat{\omega}}(d), \psi_{\hat{\mu}_b}(e)$ and $\psi_{\hat{\mu}_v}(f)$ is equal to α and other terms are ≥ α such that $\varsigma = L_1(b, d, e, f)$ to fulfill the condition:

$$\psi_{L_q}(\varsigma) = \alpha \tag{30}$$

Using the procedure described in Ref. [1], we get, the lower and upper limits of the alpha cut of the membership function corresponding to all possible cases specified in Eqn. (30) are:

- **Case 1:**

  $(L_q)_\alpha^{L_1} = \min_{\Omega} L_q(b, d, e, f)$ and $(L_q)_\alpha^{U_1} = \max_{\Omega} L_q(b, d, e, f)$
  subject to $b_\alpha^L \le b \le b_\alpha^U$, $d \in \omega(\alpha)$, $e \in \mu_b(\alpha)$ and $f \in \mu_v(\alpha)$

- **Case 2:**

  $(L_q)_\alpha^{L_2} = \min_{\Omega} L_q(b, d, e, f)$ and $(L_q)_\alpha^{U_2} = \max_{\Omega} L_q(b, d, e, f)$
  subject to $b \in \lambda(\alpha)$, $d_\alpha^L \le d \le d_\alpha^U$, $e \in \mu_b(\alpha)$ and $f \in \mu_v(\alpha)$

- **Case 3:**

  $(L_q)_\alpha^{L_3} = \min_{\Omega} L_q(b, d, e, f)$ and $(L_q)_\alpha^{U_3} = \max_{\Omega} L_q(b, d, e, f)$
  subject to $b \in \lambda(\alpha)$, $d \in \mu_b(\alpha)$, $e_\alpha^L \le e \le e_\alpha^U$, $f \in \mu_v(\alpha)$

- **Case 4:**

  $(L_q)_\alpha^{L4} = \min_{\Omega} L_q(b, d, e, f)$ and $(L_q)_\alpha^{U_4} = f_\alpha^L L_q(b, d, e, f)$
  subject to $b \in \lambda(\alpha)$, $d \in \mu_b(\alpha)$, $e \in \mu_b(\alpha)$, $f_\alpha^L \le f \le f_\alpha^U$

From the definition of $\lambda(\alpha)$, $\omega(\alpha)$, $\mu_b(\alpha)$ and $\mu_v(\alpha)$ given in Eqns. (26)–(29) $b \in \lambda(\alpha)$, $\omega \in (\alpha)$, $e \in \mu_b(\alpha)$ and $f \in \mu_v(\alpha)$ can be replaced by $b \in [b_\alpha^L, b_\alpha^U]$, $d \in [d_\alpha^L, d_\alpha^U]$, $e \in [e_\alpha^L, e_\alpha^U]$ and $f \in [f_\alpha^L, f_\alpha^U]$, respectively. We get, $(L_q)_\alpha^L = \min_{\Omega} L_q(b, d, e, f)$ and $(L_q)_\alpha^U = b_\alpha^U L_q(b, d, e, f)$ subject to $b_\alpha^L \le b \le b_\alpha^U$, $d_\alpha^L \le d \le d_\alpha^U$, $e_\alpha^L \le e \le e_\alpha^U$ and $f_\alpha^L \le f \le f_\alpha^U$.

The crisp interval $[(L_q)_\alpha^L, (L_q)_\alpha^U]$ is α-cuts of $\tilde{L}q$. Based on the convexity property and extension principle of fuzzy numbers, we get, $[(L_q)_{\alpha_1}^L, (L_q)_{\alpha_1}^U]$ is the subset of $[(L_q)_{\alpha_2}^L, (L_q)_{\alpha_2}^U]$ with the limitation $0 < \alpha_2 < \alpha_1 \le 1$, i.e., $(L_q)_\alpha^U$ decreases and $(L_q)_\alpha^L$ increases as α increases. $\psi_{\tilde{L}_q}(\varsigma)$ can be constructed, based on these bounds. If both $(L_q)_\alpha^U$ and $(L_q)_\alpha^L$ are invertible in terms of α, the M.F. $\psi_{\tilde{L}_q}(\varsigma)$ can be symbolized as $\psi_{\tilde{L}_q}(\varsigma) = \begin{cases} L(\varsigma), \varsigma_1 \le \varsigma \le \varsigma_2 \\ 1, \varsigma_2 \le \varsigma \le \varsigma_3 \\ M(\varsigma), \varsigma_3 \le \varsigma \le \varsigma_4 \end{cases}$ where left shaped function = $L(\varsigma) = [(L_q)_\alpha^L]^{-1}$ and right shaped function = $M(\varsigma) = [(L_q)_\alpha^U]^{-1}$. The Yager ranking index formula, $Y(L_q) = \frac{1}{2}\int_0^1 [(L_q)_\alpha^L + (L_q)_\alpha^U$ is used to defuzzify from fuzzy parameters to crisp one.

7.6 FUZZY NUMERICAL ANALYSIS

Let $\hat{\lambda}$ = [0.5, 1, 1.5, 2], $\hat{\omega}$ = [25, 26, 27, 28], $\hat{\mu}_b$ = [30, 31, 32, 33] and $\hat{\mu}_v$ = [10, 11, 12, 13] are the trapezoidal fuzzy numbers of the rates corresponding to arrival, retrial, and service rate in regular and vacation mode. The following parametric values are assumed for the calculation α =15, p = 0.6 and θ = 0.5. Then $[b_\alpha^L, \frac{1+\alpha}{2}] = [\frac{1+\alpha}{2}, \frac{4-\alpha}{2}]$, $[d_\alpha^L, d_\alpha^U] = [25 + \alpha, 28 - \alpha]$, $[e_\alpha^L, e_\alpha^U] = [30 + \alpha, 33 - \alpha]$ and $[f_\alpha^L, f_\alpha^U] = [30 + \alpha, 33 - \alpha]$

It is obvious that when $b = b_\alpha^U$, $d = d_\alpha^L$, $e = e_\alpha^L$ and $f = f_\alpha^L$, the mean orbit size reaches its maximum value and when $b = b_\alpha^L$, $d = d_\alpha^U$, $e = e_\alpha^U$, and $f = (L_q)_c^I$, the mean orbit size attains its minimum value. The upper and lower bounds of α-cuts of L_q are calculated by using MATLAB as:

$$(L_q)_\alpha^L = \frac{-[(a+1)(-5392800\,a^4 + 84321600\,a^3 6199012009 1\,a^2 + 4054710559 10\,a + 3290601175 89)]}{[599\,(a-33)\,(5179\,a^2 - 302990\,a + 876859)]}$$

$$(L_q)_\alpha^U = \frac{[(a-4)(5392800\,a^4 + 935208000\,a^3 + 31239236891\,a^2 + 124976996164\,a - 174105870900)]}{[599(a+30)(5179\,a^2 + 277916\,a + 305500)]}$$

Inverse function of lower and upper bounds of mean orbit size exists, which in turn gives the M.F. of the expected orbit size. In Figure 7.1, the M.F. trapezoidal in shape corresponds to the selected trapezoidal fuzzy numbers. The M.F. triangular in shape is obtained by automatic conversion from trapezoidal to triangular using MATLAB.

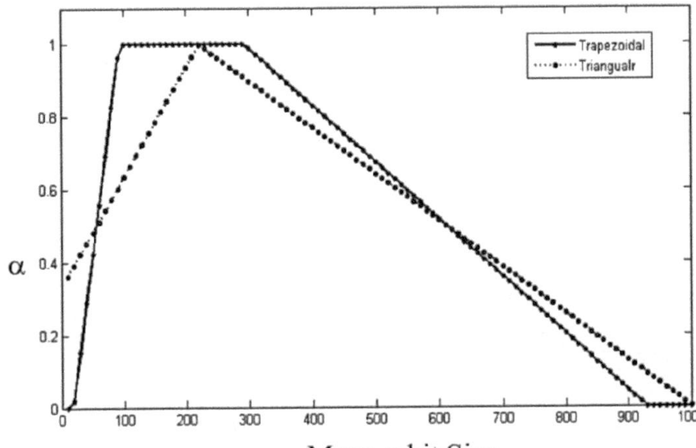

FIGURE 7.1 Comparative analysis of M.F. using triangular and trapezoidal fuzzy numbers.

Table 7.1 summarizes the mean orbit size α-cuts. The queue length range at $\alpha = 1$ is [92.7331, 292.9550]. This indicates that the expected number of customers in orbit is most likely between 92.7331 and 292.9550. Furthermore, the orbit length range for $\alpha = 0$ is [18.6875, 928.515], implying that the average number of customers in orbit will never approach 928.5151 and will never go below 18.6875.

TABLE 7.1 The α-Cut Value of Mean Orbit Size

α	$(L_q)_\alpha^L$	$(L_q)_\alpha^U$
0	18.6875	928.5151
0.1	22.9753	816.7234
0.2	27.8243	721.8064
0.3	33.2748	640.3698
0.4	39.3715	569.8745
0.5	46.1641	508.3828
0.6	53.7075	454.3905
0.7	62.0631	406.7120
0.8	71.2997	364.4004
0.9	81.4941	326.6900
1.0	92.7331	292.9550

Using Mathematica software, the defuzzification value of mean orbit size using the Yager ranking index approach is calculated, which is equal to

$$\tilde{Y}(\tilde{L}_q) = \frac{1}{2}\int_0^1 [(L_q)_\alpha^L + (L_q)_\alpha^U]\, d\alpha = 295.2929$$

The results of the numerical analysis will be extremely beneficial in constructing a queuing system that incorporates one or more combinations of numerous decisions.

7.7 CONCLUSION

The bulk arrival retrieval queue is investigated, along with Bernoulli working vacation, vacation interruption, and orbital search. The suggested model is studied in both a standard and a fuzzy context. The fuzzy queues reflect actual scenarios in conjunction with the well-known, established classical queuing systems, which are rigorous but usually fall too far short of reality. Fuzzy queues are more realistic and precise than standard queues.

KEYWORDS

- **fuzzy parameters**
- **fuzzy queue**
- **orbital search**
- **standard queues**
- **vacation interruption**
- **working vacation**

REFERENCES

1. Zadeh, L. A., (1965). *Fuzzy Sets, Information and Control, 8*, 338–353.
2. Ke, J. C., Huang, H. I., & Lin, C. H., (2007). On retrial queueing model with fuzzy parameters. *Physica A: Statistical Mechanics and its Applications, 374*, 272–280.
3. Barak, S., & Fallahnezhad, M. S., (2012). Cost analysis of fuzzy queuing systems, *International Journal of Applied Operational Research, 2*(2), 25–36.
4. Noora, A. A., Jahanshahloo, G. R., FanatiRashidi, S., & Moazami, G. M., (2011). A study of discrete-time multi server retrial queue with finite population and fuzzy parameters. *African Journal of Business Management, 5*(6), 2426–2431.
5. Gomathi, C., & Adaikalam, A., (2021). Fuzzy queue-based switch efficiency measures for optical network. *Optical and Quantum Electronics, 53*(2), 1–19. doi: 10.1007/s11082-020-02704-y.
6. Visalakshi, V., & Suvitha, V., (2018). Performance measure of fuzzy queue using pentagonal fuzzy number. *Journal of Physics: Conference Series, 1000*(1), 1–6.
7. Bagyam, J. E. A., & UdayaChandrika, K., (2019). Fuzzy analysis of bulk arrival two phase retrial queue with vacation and admission control. *The Journal of Analysis, 27*(1), 209–232. https://doi.org/10.1007/s41478-018-0118-1.
8. Kannadasan, G., & Padmavathi, V., (2021). Classical fuzzy retrial queue with working vacation using hexagonal fuzzy numbers. *Journal of Physics: Conference Series*, 2070. doi: 10.1088/1742-6596/2070/1/012018.
9. Bouchentouf, A. A., Cherfaoui, M., & Boualem, M., (2019). Performance and economic analysis of a single server feedback queueing model with vacation and impatient customers. *OPSEARCH, 56*, 300–323. https://doi.org/10.1007/s12597-019-00357-4.
10. Nayeri, S., Tavakoli, M., Tanhaeean, M., & Jolai, F., (2021). A robust fuzzy stochastic model for the responsive-resilient inventory-location problem: Comparison of metaheuristic algorithms. *Annals of Operations Research*, 1–41. https://doi.org/10.1007/s10479-021-03977-6.
11. Mishra, S. S., Singh, B. B., Yadav, S. K., Rawat, S., Mishra, P. P., & Singh, A., (2021). Cost optimization of arrival and service control queueing model with single server in fuzzy environment of uncertainty. *Turkish Journal of Computer and Mathematical Education, 8*(12), 1626–1639.

CHAPTER 8

A REVIEW ON ALUMINUM METAL MATRIX COMPOSITES: PROCESSING AND PROPERTIES

MAHENDRA KUMAR MALVIYA,[1] VANDANA SOMKUNWAR,[2] M. K. PRADHAN,[3] RAM KRISHNA RATHORE,[4] and NITISH KUMAR SINGH[5]

[1] *Department of Mechanical Engineering, Government Polytechnic College, Betul, Madhya Pradesh, India*

[2] *Department of Mechanical Engineering, NITTTR, Bhopal, Madhya Pradesh, India*

[3] *Department of Mechanical Engineering, MANIT, Bhopal, Madhya Pradesh, India*

[4] *Department of Mechanical Engineering, Rungta College of Engineering and Technology, Bhilai, Chhattisgarh, India*

[5] *School of Mechanical Engineering, VIT, Bhopal University, Bhopal, Madhya Pradesh, India*

8.1 INTRODUCTION

The composite materials are lightweight compared to other materials. Aluminum-based hybrid metal matrix nanocomposites provide properties that are not obtainable by the individual phases/metals [1, 2]. The company produces aluminum metal composites (AMC) that are important materials in aeronautical, automobile, and marine fields [3]. The properties of AMMCs include high strength and increased corrosion resistance with various percentages of

Computational Optimization, Modeling, and Simulation for Engineering Applications.
Anupam Shukla, Sourabh Rungta, Mohan Awasthy, and Rakesh Himte (Eds.)

TiB2 [5]. Al7075, SiC, and TiB2 materials were used to construct AMMCs. The Aluminum-based hybrid metal matrix nanocomposites are fabricated through the stir casting method [1, 5, 6]. Stirring the compound does not produce a bad composition in the AMMCs reinforcement. Stir-fabricated composites are low in cost and easy to fabricate. Hybrid AMMCs are widely used in several innovative engineering applications such as automobile, aerospace, and locomotive industries, etc. [8], replacing conventional metals and alloys because of their better mechanical characteristics, such as better wear resistance, small thermal coefficient, and higher specific strength [10]. Furthermore, the continuous requirement for smart materials with higher strength-to-weight ratios and cost-effective materials is growing gradually, which drives investigators to focus on numerous innovations and investigations of metal composites over the years. It has been proven that novel AMMCs exhibit better wear resistance, better density, and higher strength, which are widely used in the industry, various defense fields, structural works, and craft works [15].

8.2 ALUMINUM METAL MATRIX COMPOSITES (MMCS)

The rich availability, around 8% by mass of our earth's crust, is aluminum, making it the most useful metal and also the third useful material after silicon and oxygen [5]. Currently, lightweight materials are in higher demand in the aircraft and vehicle industries [7]. Aluminum-based hybrid metal matrix nanocomposites are particularly likely materials for those industrial purposes above regular alloys because of their superior mechanical features, such as high hardness, high wear resistance, improved temperature, and greater specific strength [5]. High stiffness, hardness, specific strength, high temperature resistance, high wear resistance, good abrasion resistance, low density, and corrosion resistance are just a few of aluminum's adaptable physical and mechanical features. They contribute to its widespread use in a variety of fields, including automobiles, aircraft, structural applications, and maritime. Few mechanical industrial uses agriculture waste like bamboo ash, fly ash, Cotton, Rice Husk Ash, coronate ash, Red Mud, Corn cob ash, Palm Kernel Shell Ash, Maize Stalk Ash, etc., the reinforced materials now a day used as reinforced material. Some of reinforcement material is – titanium diboride (TiB_2) Due to several desirable properties TiB_2 is an important material for AMCs. TiB_2 shows better refractory characteristics such as better hardness, good melting point, better modulus properties and TiB_2 reinforced with relatively high durability and strength hardness, melting point, wear resistance, and strength to density ratio ceramic material as characterized [12]. The current use of TiB_2 such as cutting tool, impact resistant armor, and wear resistance coatings. A significant application of TiB_2 composites in the wear reduction of aluminum metal, silicon carbide (SiC): Sic is most commonly used dispersion in aluminum alloys [13, 14]. The unique quality of Sic is that from all other ceramics is that it has the highest thermal conductivity (120 W/mk), hardness, and a good modulus of elasticity. It gives a homogeneous composite of aluminum. The sintered Al-Sic composites are utilized in automobile industries as pistons, cylinder liners, and brake drums [16].

8.3 PROCESSING OF ALUMINUM METAL MATRIX COMPOSITES (MMCS)

8.3.1 STIR CASTING

In stir casting, the most famous procedure to manufacture metal matrix composites (MMCs), a stirrer is used to properly mix the reinforced material, which is made of high melting temperature material. Figure 8.1 shows the general setup of stir casting. There are two main parts of the stirrer: the impeller and a rod, which is cylindrical in shape. One end of the cylindrical rod is connected to the impeller, and the other end is connected to the shaft of the motor in a vertical position. When the motor is rotated on its own axis, the rod also rotates in a vertical position [6]. While manufacturing MMCs by the stir casting method, a lot of factors need to be considered, such as the uniform distribution of reinforcement particle material, either by weight or volume, between the metal alloys and reinforcement particles, the chemical reaction between metal alloys and reinforced particles, and the bonding strength between the metal alloy and reinforcement [8].

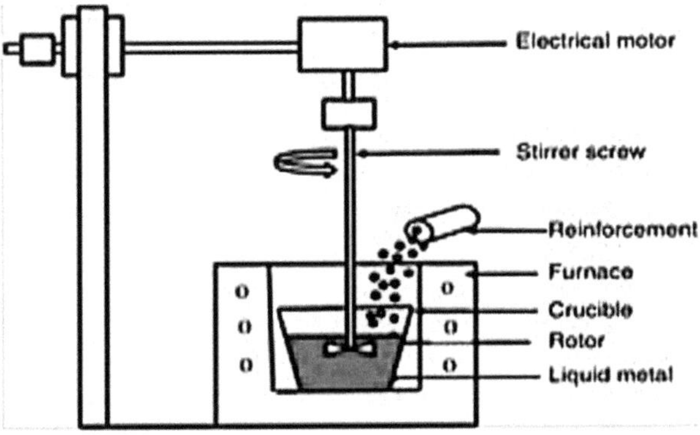

FIGURE 8.1 Stir casting setup [3].

Source: Reprinted with permission from Ref. [3].

8.3.2 *ELECTRICAL DISCHARGE MACHINING (EDM)*

MMCs have a strong reinforcement that makes machining methods challenging. Tool wear and reduced tool life are caused by hard and abrasive reinforcing particles [12, 13]. Advanced non-conventional machining processes such as abrasive jet machining, laser beam machining (LBM), and water jet machining (WJM) are successfully employed, however the surface finish is poor. The cost of equipment and the height of the workpiece are two key factors in LBM, AJM, and WJM machining methods [19]. EDM (electro discharge machine) and WEDM (wire electric discharge machining) are examples of successful processes that produce satisfactory results. Both the workpiece and the electrode must be electrically conductors in EDM. The reshaped or created electrode (tool) is shaped to the type of hole it is to repeat and is usually made of graphite or copper. The hard reinforcement available in MMCs creates problems for machining methods. Hard and abrasive reinforcing particles are the reason for tool wear, which shortens the tool life [12, 13]. Advanced non-conventional machining processes like abrasive jet machining, LBM process, and WJM process are utilized effectively but result in bad surface finish. In LBM, AJM, and WJM machining methods, equipment cost and height of the workpiece are two major factors [19]. Processes such as EDM and WEDM are very efficient and provide acceptable outcomes. In EDM, both the workpiece and the electrode must be conductors of electricity. The reshaped or formed electrode (tool), normally produced using graphite or copper, is shaped to the type of hole it is to repeat.

8.4 CHARACTERISTICS OF AMMCS (TABLE 8.1)

TABLE 8.1 Overview of Matrix, Method, and Reinforcement Properties Investigated for Aluminum Metal Matrix Composites

Sl. No.	Year/Author(s)	Material Used	Method Used	Reinforce Material Used	Output	Remarks
1.	Rao et al. [1]	AA6061	Friction stir method	TiB_2	Sliding wear by using pin disc method. Corrosion behavior by salt spray technique.	Other characterization missing.
2.	James et al. [2]	Al6061	Stir casting bottom pouring technique.	$SiC/Al_2O_3/TiB_2$	Strength and property Micro-structural (SEM)	Other characterization missing.
3.	Kumar et al. [3]	Al7075	Friction stirs processing (FSP)	2 wt.% SiC	Tensile and compressive strength, ductility, hardness.	Variation of wt.% of reinforce material missing.

TABLE 8.1 *(Continued)*

Sl. No.	Year/Author(s)	Material Used	Method Used	Reinforce Material Used	Output	Remarks
4.	Dakarapu S R et al. [4]	AA6061	Friction stirs casting process.	TiB_2	Tensile strength and hardness are optimized by using the GRA method, which considers axial load, rotational speed, transverse feed, and reinforcement percentage as input parameters.	–
5.	Mallikarjuna et al. [5]	Aluminum powder	Stir casting technique.	Fly ash (2%, 4%, 6% by weight) and S-glass fibers (3%, 5%, 7% by wt.).	Impact strength	Microstructure testing, tensile strength, compressive strength., hardness, etc., missing.
6.	Daniel et al. [6]	Al	NA	SiC (5, 10%, 15% by weight) and particle size (10, 20, 40 μm), MoS_2 2%).	Wear rate and coefficient of friction (pin disc method).	Fabrication and other properties missing.
7.	Singh et al. [7]	Al powder	Powder metallurgy method.	SiC, Al_2O_3, WS_2	Wear rate and coefficient of friction (pin disc method) and optimize by ANOVA technique.	Optimizations vie modern technique is missing.
8.	Ghasali et al. [8]	Aluminum powders (pure and 1056)	Spark plasma and sintering method.	Al-SiC-TiC	Thermal expansion.	Optimization of process parameter and other characterization is missing.
9.	Singh et al. [9]	Al	Processing route	SiC and graphite.	Density	Fabrication is done by old method and another characteristic is missing.
10.	Narasaraju et al. [10]	AlSi 10 Mg	Stir casting.	Fly ash (15%) + rice husk ash (5%)	Ductility (decrease when we increase the weight percentage of fly ash).	Optimization of process parameter is missing.
11.	Eskandari et al. [11]	Al alloys	Friction stirs processing.	TiB_2/Al_2O_3	Hardness	Optimization and machining parameter calculation missing.
12.	Bodunrin et al. [12]	Al6063	–	SiC, rice husk	Tensile strength, fracture toughness.	Fabrication process missing.
13.	Fatile et al. [13]	Al-Mg-Si alloy	Two-step stir casting.	SiC, corn cob ash.	Hardness, ultimate tensile strength, and percentage of elongation is inversely proportional to the weight of corn cob ash.	Optimization of process parameter is missing.
14.	Baskaran et al. [14]	Al7075	Liquid metallurgy route.	TiC, Al_2O_3	Micro-structural studies wear behavior (coefficient of friction and wear increase with the decrease the percentage of TiC.	Optimization of process parameter missing.

TABLE 8.1 *(Continued)*

Sl. No.	Year/Author(s)	Material Used	Method Used	Reinforce Material Used	Output	Remarks
15.	Show et al. [15]	635 Al alloy.	Stir casting process.	With 4% (Al_2O_3+ SiC)	Wear behavior.	Other mechanical properties and optimization of process parameter missing.
16.	Uthaya Kumar et al. [16]	Al alloy	Stir casting process.	SiC and B_4C (5, 10, 15% by weight).	Coefficient of friction, wear resistance.	Other properties such as microstructure, strength are missing.
17.	Boopathi et al. [17]	Al alloys	Stir casting	SiC, fly ash (15% by weight).	Density	Other properties such as microstructure, strength are missing.
18.	Alaneme et al. [18]	Al-Mg-Si alloy.	Double stir casting.	Rice husk ash (RHA) and alumina (Al_2O_3).	Coefficient of friction, wear resistance.	Other properties such as microstructure, strength are missing. Optimization of machining process parameter is missing.
19.	Nagaral et al. [18]	Al6061	Stir casting	Al_2O_3(3, 6, and 9%).	Hardness, tensile, and yield strength.	Microstructure structure missing.

8.5 CONCLUSIONS

Using the literature survey, it has been clearly identified that aluminum alloy based MMCs can be successfully produced by stir casting with a uniform distribution of silicon carbide and TiB_2 particles. It is found from the experiment that the tensile strength and the compressive strength increase while increasing the reinforcement particles in the MMCs. Atomic force microscopy (AFM) analysis microstructure was visualized with the help of AFM. Scanning electron microscope (SEM) images Show that SiC and TiB_2 particulates are uniformly distributed in the aluminum alloy matrix. Increased reinforcing particles such as TiB_2 and SiC increased mechanical characteristics. It is concluded that TiB_2 and SiC particle mixed uniformly in stir casting. For the literature review, it is clear that most researchers worked on the effect of machining process parameters of EDM machines on composite materials. In the literature review, silicon carbide is mostly used as a reinforced composite for the manufacturing of aluminum composites. Most researchers have studied the micro-machining behavior, and some researchers have investigated the study of hybrid MMCs. Therefore, our future investigation of Al/ TiB_2 /SiC AMCs will focus on EDM studies on MMCs.

AMCs are the most effective way to increase mechanical property and are mostly used in aerospace industries and automobiles nowadays. From a review of the literature, it appears that the reinforcement of titanium diboride (TiB_2)

is less common, but it has a lot of potential due to its high strength, hardness, and wear and corrosion resistance. To optimize the machining settings, experiments are devised and executed on a Wire Electrical Discharge Machine with AA7075/TiB_2 (0, 3, 6 wt.%) as the work material. The answers are the MRR and surface roughness. The proposed Grey-based Taguchi approach improves multi-response optimization. No one has prepared/attempted AA7075/TiB_2/SiC hybrid metal composite and tested mechanical behavior AMCs are the most effective way to increase mechanical property and are mostly used in aerospace industries and automobiles nowadays. Following are the major research gaps identified from the overview of past literature: (i) To prepare AA7075/TiB_2/SiC (aluminum alloy composites) using stir casting by varying weight percentage of TiB_2 and SiC; (ii) investigation of machinability parameters of hybrid AMMC using EDM; (iii) investigation of the mechanical property of hybrid AMMC and comparison in different wt.% of ceramics in AMCs. Optimize the performance parameters using grey based Taguchi method; and (iv) compare the optimized experimental results available in the literature with optimized results.

KEYWORDS

- **aluminum metal matrix composites**
- **atomic force microscopy analysis microstructure**
- **electrical discharge machining**
- **hybrid metal composite**
- **silicon carbide**
- **stir casting**
- **Taguchi method**

REFERENCES

1. Rao, S. D., & Ramanaiah, N., (2017). Evaluation of wear and corrosion properties of AA6061/TiB_2 composites produced by FSP technique. *Journal of Minerals Characterization and Engineering, 5,* 353–361. http://www.scirp.org/journal/jmmce,ISSNOnline: 2327-4085.
2. James, S., Annamalai, A., Kuppan, P., & Oyyaravelu, R., (2017). Fabrication of hybrid metal matrix composite reinforced with SiC/Al_2O_3/TiB_2. *Mechanics, Materials Science & Engineering MMSE Journal.* Open Access. https://hal.archives-ouvertes.fr/hal-01504690 (accessed on 21 December 2023).
3. Kumar, A., Pal, K., & Mula, S., (2017). Simultaneous improvement of mechanical strength, ductility, and corrosion resistance of stir cast Al7075-2% SiC micro-and nano composites by friction stir processing. *Journal of Manufacturing Processes, 30,* 1–13.

4. Santha, R. D., & Ramanaiah, N., (2017). Process parameters optimization for producing AA6061/TiB_2 composites by friction stir processing. *Journal of Mechanical Engineering – Strojnícky Časopis, 67*(1), 101–118.
5. Ch. Mohana, R., & Mallikarjuna, R. K., (2016). *Abrasive Wear Behavior of TiB_2 Fabricated Aluminum 6061.* PMME 2016 Elsevier Ltd. All rights reserved. Selection and Peer-review under responsibility of International Conference on Processing of Materials, Minerals and Energy, Ongole, Andhra Pradesh, India.
6. Daniel, A. A., Murugesan, S., Sukkasamy, S., et al., (2017). Dry sliding wear behavior of aluminum 5059/SiC/MoS_2 hybrid metal matrix composites. *Materials Research, 20*(6), 1697–1706.
7. Singh, N., Mir, I. U. H., Raina, A., Anand, A., Kumar, V., & Sharma, S. M., (2017). Synthesis and tribological investigation of Al-SIC based nano hybrid composite. *Alexandria Engineering Journal.*
8. Ghasali, E., Yazdani-rad, R., Asadian, K., & Ebadzadeh, T., (2017). Production of Al-SIC-TIC hybrid composites using pure and 1056 Aluminum powders prepared through microwave and conventional heating methods. *Journal of Alloys and Compounds, 690*, 512–518.
9. Singh, J., (2016). Fabrication characteristics and tribological behavior of Al/SIC/Gr hybrid aluminum matrix composites: A review. *Friction, 4*(3), 191–207.
10. Narasaraju, G., & Raju, D. L., (2015). Characterization of hybrid rice husk and fly ash-reinforced aluminum alloy (Alsi10mg) composites. *Materials Today: Proceedings, 2*(4, 5), 3056–3064.
11. Eskandari, H., & Taheri, R., (2015). A novel technique for development of aluminum alloy matrix/TIB2/AL_2O_3 hybrid surface nano composite by friction stir processing. *Procedia Materials Science, 11*, 503–508.
12. Bodunrin, M. O., Alaneme, K. K., & Chown, L. H., (2015). Aluminum matrix hybrid composites: A review of reinforcement philosophies; mechanical, corrosion and tribological characteristics. *Journal of Materials Research and Technology, 4*(4), 434–445.
13. Fatile, O. B., Akinruli, J. I., & Amori, A. A., (2014). Microstructure and mechanical behavior of stir-cast Al-Mg-Sl alloy matrix hybrid composite reinforced with corn cob ash and silicon carbide. *International Journal of Engineering and Technology Innovation, 4*(4), 251–259.
14. Baskaran, S., Anandakrishnan, V., & Muthukannan, D., (2014). Investigation on dry sliding wear behavior of in-situ casted AA7075-TiC metal matrix composites by using Taguchi technique. *Mater. & Dsg., 60*, 186–192.
15. Bijay, K. S., Dipak, K. M., & Joydeep, M., (2014). Dry sliding wear behavior of aluminum-based metal matrix composites with single (Al_2O_3) and hybrid (Al_2O_3 + SiC). *Metallogr. Microstruct. Anal., 3*, 11–29.
16. Uthaya, K. M., Aravindan, S., & Rajkumar, K., (2013). Wear performance of Al–SiC–B_4C hybrid composites under dry sliding conditions. *Mater. & Dsg., 47*, 456–464.
17. Mahendra, B. M., Arulshri, K. P., & Iyandurai, N., (2013). Evaluation of mechanical properties of aluminum alloy 2024 reinforced with silicon carbide and fly ash hybrid metal matrix composites. *American J. App. Sci., 10*, 219–2295.
18. Altinkok, N., Ozsert, I., & Findik, F., (2013). *Dry Sliding Behavior of Al_2O_3/SiC Particle Reinforced Aluminum Based MMCs Fabricated by Stir Casting Method, 124*, 11–19.
19. Madeva, N., Bharath, V., & Auradi, V., (2013). Effect of Al_2O_3 particles on mechanical and wear properties of 6061 Al. alloy metal matrix composites. *Mater. Sci.& Engg., 2*, 117–124.

CHAPTER 9

MACHINING INVESTIGATION OF AL7075/AL2O3 COMPOSITES FABRICATED VIA STIR CASTING TECHNIQUE

SHAIK IQBAL ALI BASHA, RAM KRISHNA RATHORE, PEEYUSH DEWANGAN, SURAJ BANDHEKAR, TEJASVI ANANT, and ASHOK KUMAR SHUKLA

Department of Mechanical Engineering, Rungta College of Engineering and Technology, Bhilai, Chhattisgarh, India

ABSTRACT

Aluminum (Al) is widely used as a base matrix in metal matrix composites (MMCs) due to its favorable properties. Al-MMCs, particularly Al7075/Al_2O_3, show promise for applications in automotive and aerospace industries due to their lightweight nature and exceptional mechanical properties. Previous research emphasizes the importance of parameters like feed rate (FR) and depth of cut (DOC) in achieving optimal surface roughness (SR) and minimizing tool wear. Al_2O_3 nanoparticles are used as reinforcement, and Al6061 as the matrix. The squeeze casting technique is employed, ensuring proper mixing and reduced defects in the fabricated composites. Microscopic analysis is conducted to examine grain morphology. Machining operations are performed under different conditions, analyzing parameters like feed rate, cutting speed, and machining environment (dry vs. minimum quantity lubrication – MQL). Surface profilometer measurements reveal the impact of these parameters on surface roughness. The study observes that Al_2O_3 reinforcement leads to a finer grain structure, affecting hardness and

Computational Optimization, Modeling, and Simulation for Engineering Applications.
Anupam Shukla, Sourabh Rungta, Mohan Awasthy, and Rakesh Himte (Eds.)

brittleness. Cutting forces increase with higher FR but are lower in MQL machining due to lubrication effects. Surface roughness tends to decrease with increased cutting speed. Squeeze casting followed by machining operations demonstrates the feasibility of producing Al-based composites with improved properties. MQL machining shows advantages over dry machining, resulting in smoother surfaces and reduced cutting forces.

9.1 INTRODUCTION

In today's world of quickly advancing technology, engineers and researchers are designing new, stronger, and lighter materials that may be employed in a variety of industries to increase their productivity. Aluminum (Al) is the commonly used metallic alloy as base matrix element for creation of metal matrix composites (MMCs) owing to its superior characteristics (thermal, electrical, corrosion, and mechanical), light in weight, ease of availability and machinability. In the available literature, Al-MMCs is found as the most promising material for fabricating automobile components due to their low weight and exceptional mechanical properties [1]. They are comparable to cast iron in terms of strength and wear resistance, have three times the heat conductivity, and are approximately 67% less dense. Owing to the above-mentioned properties, Al-MMC alloys are a promising candidate for the fabrication of lightweight aircraft and automobile components [2]. Earlier literature shows that for preparing Al-MMCs, melt stirring processing routes are widely used [3, 4]. Bhushan et al. [5] found out that at the feed rate (FR) range of 0.1–0.3 mm/rev and depth of cut (DOC) of the range 0.5–1.5 the surface roughness of the Al7075/10 wt.% SiC was observed to be the minimum in the experiment. They also examined the wear of the tool and observed that the wear of the tungsten carbide cutting tool enhances by 2.4 factor with varying the cutting speed from 180 to 240 meter per minute at 0.5 mm and 0.1 mm/rev DOC and FR, respectively. Researchers [6] also performed an investigation to analyze the influence of angle of approach ranging from 45?–90?, 0.05 to 0.2 mm/rev FR and cutting speed ranging from 80 to 170 m/min on the SR employing surface roughness tester. They observed lower SR while machining hybrid composite specimens over single-reinforced hard composites. They also observed that by adding self-lubricant solid nanoparticles in the composites, lower SR can be achieved. Soni & Thomas [7] performed the comparative analysis of electrochemical machining process parameters employing the Taguchi and Genetic Algorithm approach. To bridge the gaps in the literature, a study was conducted to perform different operations which focus on the development of a new composite with Al as the base matrix, and its various properties which make them better than the existing metals or materials [8, 9].

9.2 MATERIAL AND METHODS

The Al_2O_3 nanoparticles were procured from local vendors with an average size of 40–60 nm and employed as a reinforcement. Al6061 was purchased from the local vendors and taken as a matrix material. Squeeze casting is used for the preparation of Al-based MMC. Before starting the process, the furnace and the die were cleaned thoroughly and a single layer of non-stick coating of graphite paste is applied to the furnace and the die. The furnace temperature is allowed to reach 950°C and meanwhile, the die is preheated separately at 250°C to prevent shrinkage and reduce blowholes. The lower duct of the furnace is unclogged once the furnace has reached the desired temperature. This is an important step as the clogged impurities from the previous casting process will not let the molten metal flow properly into the die. Once the unclogging is done properly, the casting machine is ready for the fabrication of the first sample. For the sample preparation, 1.2 kg of Al6061 blocks are placed in the furnace to melt and at the same time *SiC* particles are preheated at 250°C. After every 10 minutes, the lid of the furnace is opened to check if the Al blocks are melted. Once the Al blocks are melted completely, the slag formed on the molten aluminum upper surface is removed and a teaspoon of coverall is added to increase the wettability, remove impurities, and reduce porosity. The preheated powder mixture is added to the molten Al inside the furnace with the help of the funnel. The mixture is then stirred for 10 minutes at a stirring speed of 550 rpm to assure proper mixing. The preheated die is kept below the duct of the furnace, and the mixture is poured from the duct into the die with the help of a heated funnel. The molten metal in the die is squeezed under the pressure of 150 MPa until the die is cooled and solidified. The stirrer is moved away from the furnace and the remaining slag is removed from the furnace. Once the die is cooled, the sample is taken out from the die and machining is performed on that sample. After the preparation of the specimen, microscopic analysis has also been performed to analyze the grain morphology of the fabricated composite.

9.3 MACHINING ANALYSIS

At first, the irregular portion of the fabricated composite cylindrical rod sample is machined employing skinning operation in a lathe machine. Then, the fabricated specimen was fitted in a Computer Numerical Control (CNC) system to evaluate the parameters' effect, i.e., DOC at 0.5 mm, CS as 200 and 180 m/min, and feed rate as 0.3, 0.2, 0.1 mm/rev, and on the surface profiles under minimum quantity lubrication (MQL) and dry turning environments. Dry machining was completed first, followed by MQL. At room temperature, all of the experiments were carried out. All of the tests were carried out at ambient conditions. A surface profilometer was used to depict the surface profile of each of the machined surfaces. The SR was measured in multiple locations on the fabricated cylindrical specimen, and the mean value was taken as a final reading (Table 9.1).

TABLE 9.1 Parameters for Conducting Experiments

Exp. Run	FR (mm/rev)	DOC (mm)	CS (m/s)	Material	Machining Environment	Output Variables
1	0.1	0.5	180	Al6061 + Al_2O_3	Dry	Surface roughness
2		0.5	200			
3	0.2	0.5	180			
4		0.5	200			
5	0.3	0.5	180			
6		0.5	200			
7	0.1	0.5	180	–	MQL	–
8		0.5	200			
9	0.2	0.5	180			
10		0.5	200			
11	0.3	0.5	180			
12		0.5	200			

9.4 RESULTS AND DISCUSSION

The composite with Al_2O_3 is the finest structure grain structure which is why it acts as a hard material and is very brittle. Figure 9.1 shows the microstructures of the of Al7075/Al_2O_3 composite specimen.

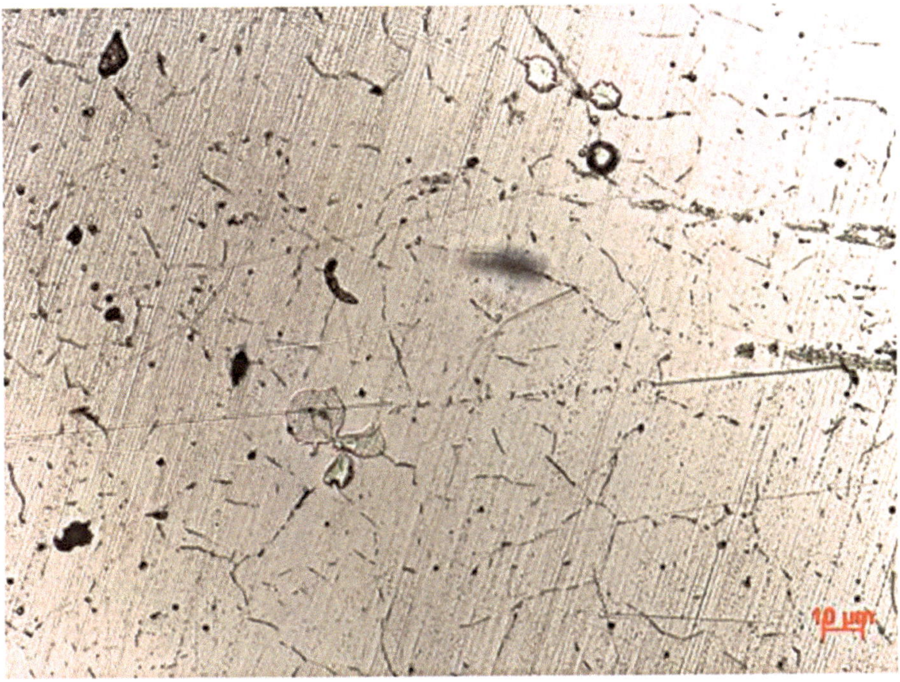

FIGURE 9.1 Optical microscopy image of the $Al7075/Al_2O_3$ composite specimen.

The forces that were established during the machining of the squeeze cast Al-based composites for different FRs and cutting speed were recorded and are shown in Figure 9.2. It was discovered that as the FR enhances, the cutting force increased as well. The area contact of the work-tool grows as the FR increases, raising the interface temperature. During the machining of the unreinforced aluminum alloy, an increase in interface temperature creates BUE on the tool's surface. In light of this, it is reasonable to predict that cutting force will grow. Cutting forces were found to be lower for MQL machining of Al-based composites than for dry machining. The film produced by MQL (Rehbinder effect) reduces the energy formation of the material, which increases elastic-plastic deformation and decreases the cutting force. The increased speed in MQL helped to improve droplet penetration, which in turn helped to improve wetting. The cooling and lubricating action of MQL results in a reduction in cutting force. Under the MQL scenario, there was a considerable drop in feed force. This may be because of the lubrication effect of MQL.

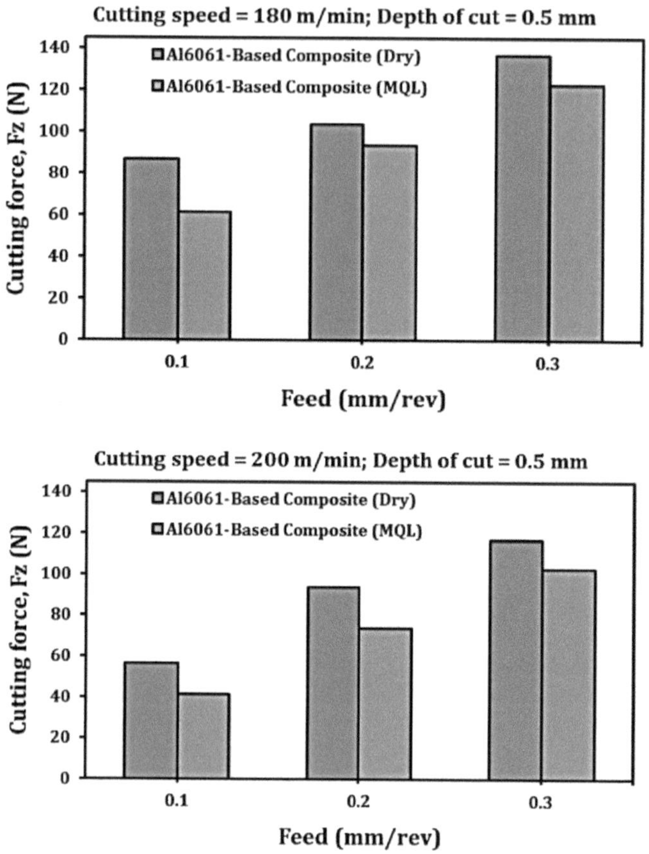

FIGURE 9.2 Comparison of cutting forces at cutting speed (a) 180; and (b) 200 m/min.

Figure 9.3 shows the comparative analysis of SR data at different machining conditions for all the prepared samples. Theoretically, the surface roughness can be calculated using the equation:

$$Ra \approx \frac{0.032 f^2}{r\varepsilon}$$

where; $r\varepsilon$ is the nose radius; f is the feed; and *Ra* is the arithmetic mean value of surface roughness. It is clear from the equation above that surface roughness rises as FR rises. Similarly, the machining experiments showed the same trend for all three samples. It was also observed that the surface becomes smoother as the cutting speed is increased. Built-up edges (BUE) are formed at low cutting speeds, and the chip gets fractured easily. They might obstruct the process of cutting, thereby surface roughness increases. However, when speed rises, the BUE lowers and breakage of chip diminishes, resulting in a smooth surface. It was discovered that the surface roughness (SR) increased as the feed rate (FR) increased. At lower feed rates, the surface profile under dry machining was discovered to be smoother. In general, with an increase in feed rate, the temperature and pressures rise, and the amount of material removed increases. This leads to increased cutting tool wear, which in turn increases surface roughness. Because Minimum Quantity Lubrication (MQL) fluid has a lubricant effect, which reduces the amount of friction at the chip-tool contact over-rake surface, MQL machining exhibits higher surface quality in contrast to dry machining.

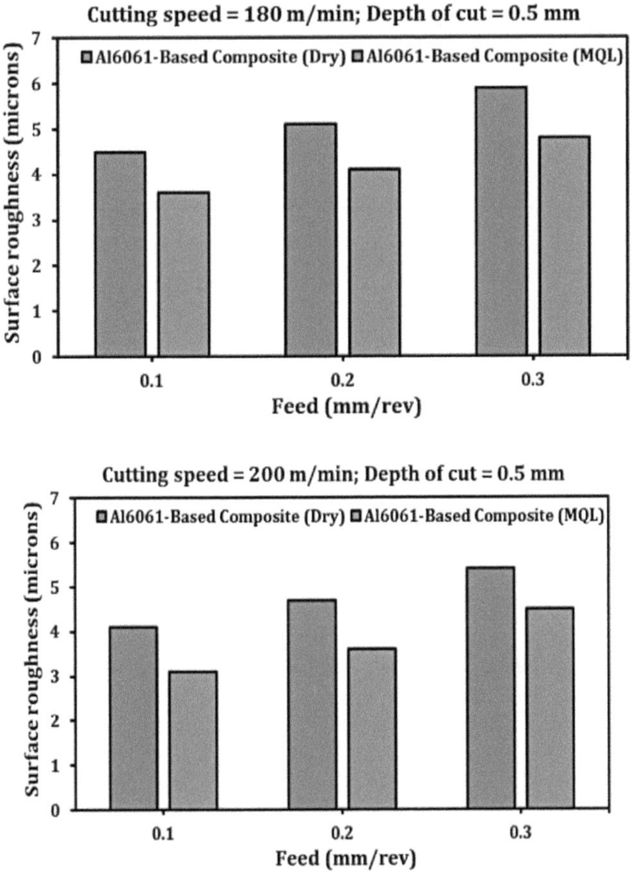

FIGURE 9.3 Comparison of the roughness of Al-based composites under dry and MQL machining.

9.5 CONCLUSION

The study focuses on the preparation of Al-based composites by the process of squeeze casting, then performing a microstructure analysis and machining operation on them under different machining conditions. The method of squeeze casting was chosen because it was known to reduce any irregularity in the casting produced. Different feed rates (FR) and cutting speeds (CS) were used for the turning process, which was done under both minimum quantity lubrication (MQL) and dry-cutting environments. A substantial decrease in feed force was evident when subjected to MQL machining environments. It is observed that the surface becomes smoother as the cutting speed is increased. Moreover, as the feed rate increases, the surface roughness (SR) is enhanced. It was found that the output was more ideal while performing MQL machining than dry machining under all varying parameters.

KEYWORDS

- **Al2O3**
- **Al6061**
- **composites**
- **machining**
- **minimum quantity lubrication**
- **squeeze casting**

REFERENCES

1. Hariharasakthisudhan, P., Rajan, B. S., &Sathickbasha, K., (2020). Inspiration of reinforcements, manufacturing methods, and microstructural changes on wear behavior of metal matrix composites—A recent review. *Materials Research Express*, *7*(1), 012006.
2. Kumar, V. M., & Venkatesh, C. V., (2019). A comprehensive review on material selection, processing, characterization, and applications of aluminum metal matrix composites. *Materials Research Express*, *6*(7), 072001.
3. Ramanathan, A., Krishnan, P. K., &Muraliraja, R., (2019). A review on the production of metal matrix composites through stir casting–furnace design, properties, challenges, and research opportunities. *Journal of Manufacturing Processes*, *42*, 213–245.

4. Srivyas, P. D., &Charoo, M. S., (2018). Role of fabrication route on the mechanical and tribological behavior of aluminum metal matrix composites: A review. *Materials Today: Proceedings*, *5*(9), 20054–20069.
5. Bhushan, R. K., Kumar, S., & Das, S., (2010). Effect of machining parameters on surface roughness and tool wear for 7075 Al alloy SiC composite. *The International Journal of Advanced Manufacturing Technology*, *50*(5), 459–469.
6. Kumar, R., & Chauhan, S., (2015). Study on surface roughness measurement for turning of Al 7075/10/SiCp and Al 7075 hybrid composites by using response surface methodology (RSM) and artificial neural networking (ANN). *Measurement*, *65*, 166–180.
7. Soni, S. K., & Thomas, B., (2017). A comparative study of electrochemical machining process parameters by using GA and Taguchi method. In: *IOP Conference Series: Materials Science and Engineering* (Vol. 263, No. 6, p. 062038). IOP Publishing.
8. Ram, K. R., Nitish, K. S., Agnivesh, K. S., Sanjay, K. P., & Ashok, K. S., (2022). Mechanical properties of lightweight aluminum hybrid composite foams (AHCFs) for structural applications.*Advances in Materials and Processing Technologies*.doi: 10.1080/2374068X.2022.2048498.
9. Rathore, R. K., Singh, N. K., & Xavier, J. F., (2021). Characterization of AA7075 alloy foam using calcium and magnesium carbonate as foaming agent. In: Pal, S., Roy, D., & Sinha, S. K., (eds.), *Processing and Characterization of Materials; Springer Proceedings in Materials,* (Vol. 13). Springer, Singapore. https://doi.org/10.1007/978–981-16-3937-1_30.

CHAPTER 10

A REVIEW ON DRUG DELIVERY SYSTEM

PARVEEN NEHRA, RITIKA SINGH, and RINKEE VERMA

Assistant Professor, Rungta College of Pharmaceutical Science and Research, Raipur, Chhattisgarh, India

ABSTRACT

Drug delivery is the process of dispensing medicated drug molecules to achieve a beneficial effect in humans and animals. For the treatment of various human disorders, nasal and respiratory routes of drug delivery systems (DDS) are achieving popularity. These routes are used in place of intravenous, intramuscular (IM), subcutaneous drug delivery, especially for polypeptide and protein medication. For this study, different types of targeted delivery systems are formulated and being analyzed for nasal route and pulmonary route of drug delivery. This route comprises different drug delivery formulations such as proliposomes, liposomes, microspheres, gels, among other delivery systems. Nanoparticles containing natural and artificial polymers show affirmation of the strict essential conditions placed on different delivery systems, likely capability to be transmitted into an aerosol, stability against different types of forces created during the aerosolization process, biocompatibility, different sites or particular cell targeting, tissues in the lung, drug release in a fixed way, and destruction within a suitable period of time interval.

10.1 INTRODUCTION

Drug delivery systems (DDS) are used to transport different types of drugs in the body in order to achieve the desired beneficial effect. These systems

Computational Optimization, Modeling, and Simulation for Engineering Applications.
Anupam Shukla, Sourabh Rungta, Mohan Awasthy, and Rakesh Himte (Eds.)

are designed to (i) enhance water solubility and chemical stability of the active drug molecule, (ii) enhance therapeutic activity, and (iii) decrease toxic effects in the human body. The main aim of a drug delivery system is to contribute to and sustain different concentrations of the drug at various target sites. Nowadays, in DDS, nanoparticles have been shown to have great possibilities as conveyer molecules in recent years. Incorporation of drug molecules in nanoparticles, such as micelles, liposomes, nanocapsules, and nanospheres, enhances the therapeutic index and decreases various toxic effects. For example, liposomal DDS can enhance bioavailability, increase effectiveness, and decrease side effects [1]. From drug carrier molecules, we use soluble polymers, micro-particles made of biodegradable, natural, and synthetic polymers, microcapsules, lipoproteins, liposomes, and micelles. The carriers are formulated in such a way that they degrade or release the drug slowly, are stimulus reactive (e.g., temperature-sensitive), and targeted (e.g., by combining them with specific antibodies against certain types of cells). Targeting is the ability to deliver the drug-loaded system to different parts of the body. Two mechanisms can be differentiated for delivering the drug to the desired sites for drug release:

i. Passive targeting; and
ii. Active targeting.

A delivery system is comprising of:

- Drug formulation;
- Dosage form;
- Mechanism for the release of drug in body.

These systems distinguish as drug release systems are in a controlled manner and target drug on a particular site.

The medicated benefits of new systems are:

- Increased effectiveness of the drug;
- Particular organ drug delivery;
- Reduce toxic effects;
- Increased usefulness;
- Better calming effect of patient.

There is no uniform and established definition of DDS. Drug delivery system based on two basic parameters:

i. Route of entry; and
ii. Dosage form [2].

10.2 LIPOSOME AND DRUG DELIVERY SYSTEM

Liposomes were first introduced by a scientist named Bangham in 1965. He found that liposomes are vesicular structures consisting of hydrated bilayers in which phospholipids are dispersed in water. Liposomes are defined as bilayer structures in which the lipid bilayer is separated from the aqueous phase. Phospholipids are the main components of naturally occurring bilayers. These phospholipids include phosphatidylcholine (PC), phosphatidylethanolamines (PE), and phosphatidylserines (PS). The main characteristic that bilayer-forming compounds share is their amphiphilicity, meaning they have a polar or aqueous head and non-polar or lipid-soluble tail regions. The main antitumor drug, doxorubicin, was formulated as a liposome in the 1980s to improve its therapeutic index. There are many mechanisms by which liposomes act within and outside the body, which are as follows:

- Liposomes attach to the cell membrane and appear to fuse with the cell membrane, releasing their content into the cell.
- When they are in the cell and their phospholipid molecules are integrated into the cell membrane, the drug is released, which is trapped inside the cell.
- When scavenger cells take liposomes inside, the active pharmaceutical ingredients (APIs) are released.

Biological properties of liposomes:

- Liposomes are biocompatible.
- Liposomes can entrap hydrophilic pharmaceutical agents inside the water compartment and hydrophobic pharmaceutical agents into the cell membrane.
- Liposome-incorporated pharmaceuticals do not cause unwanted side effects.

Advantages of liposomes:

- Drug released in a controlled manner;
- Sustainable, biocompatible, adjustable;
- Non ionized;
- Sustained release provide;
- Site specific drug delivery;
- Therapeutic index of drugs is increased [3].

10.3 LUNG-SPECIFIC DRUG DELIVERY SYSTEM

The respiratory system in the human body is a complex organ system. The lung-specific delivery system consists of two parts: airways conducting and the respiratory part. The airway region is again divided into the nasal cavity, which contains sinuses, the oropharynx, larynx, bronchi, nasopharynx, bronchioles, and trachea. The respiratory part contains respiratory bronchioles, alveolar sacs, and alveolar ducts. Substances that are inhaled deposit on a layer called mucin, which lines the conducting airways region walls. Mucus is secreted by goblet cells, with mucin being the main component that forms a gel-like film. Ciliated cells move the mucus upwards in the lung and outside of the lung, clearing allergy-causing particles.

10.3.1 MECHANISMS AND DIFFERENT WAYS OF PULMONARY DRUG ADMINISTRATION

From the previous few years, the absorption of medicated agents in blood after respiratory application has been indicated in animals and humans. By the pulmonary route, the medication is distributed through two main routes: through occurring within the nose means infusion through the nasal cavity, in which the airway lumen is narrow, and second, inhalation through the oral route. By inhalation through the oral route, for best results can be predicted as it permits only fine molecules with a varying concentration range loss of only 20% when compared with 85% by the intranasal route. This is a quick and reliable method of drug delivery to the pulmonary part. This process is very simple, cost-effective, and non-uniform drug distribution of drug.

10.4 BRAIN TARGETED DRUG DELIVERY

In brain drug development, the rate-limiting role of the BBB:

- Blood-brain barrier is the major obstacle toward brain-targeted drug delivery.
- BBB has the systematic ability to separate the human brain from the blood circulatory system, and only allows the transportation of molecules that play a vital role in the functional activity of the brain.
- It limits the transport of water and hydrophobic substances from blood circulation into the brain.

- Different receptors, enzymes, and transport systems have been recognized in the endothelium of the BBB that restrain the molecules infiltration, for example proteins and peptides are transported by receptor-mediated transcytosis [5].

10.4.1 BLOOD-BRAIN BARRIER (BBB) DISRUPTION IN SOME DISEASED CONDITIONS

The blood-brain barrier (BBB) is obstructed under various diseased conditions such as brain stroke, diabetes, different types of seizures, hypertensive encephalopathy, AIDS, various traumatic brain injuries, multiple sclerosis, Parkinson's disease (PD), and Alzheimer's disease (AD). In certain diseased conditions, remodeling protein complex which is in inter-endothelial region is the main reason for the BBB disruption [6]. Albumin, a large molecule of protein which is used as an indicator for studying leakage condition in the blood-brain barrier, barely crosses the healthy BBB. In multiple sclerosis, a diseased condition, depletion of organization at junctional sites in cholesterol-rich cell membrane regions in the brain contributes to the increased BBB permeability [7].

10.4.2 BLOOD-BRAIN TUMOR BARRIER (BBTB)

Glioma is the most common primary tumor of the brain. To match the quick growth of the tumor and migration, brain tumor cells favor the BBB at early stages. The BBB is damaged when the tumor reaches a diseased level. As a result, new blood vessels form the blood-brain tumor barrier (BBTB) and this BBTB is different from the BBB. One must target tumor cells or glioma cells when passing through the BBTB. Coating the cell with permeable proteins on the surface of nanoparticles could work for this purpose.

10.4.2.1 NANOTECHNOLOGY

10.4.2.1.1 Brain Delivery of Drugs Through Nanoparticulate System

Delivering drugs to the brain through nanoparticles is another possibility. Nanoparticles are particles of polymer which consist of natural or artificial polymers with a size range between approximately 10–1,000 nm (i.e., 1 mm). Only poly(butyl cyanoacrylate) nanoparticles are the nanoparticles that are used effectively for *in vivo* drug delivery to the brain.

The nanotechnology includes following types of nanoparticles:

- Coated nanoparticles;
- Pegylated nanoparticles;
- Solid lipid nanoparticles (SLN);
- Nanogels.

10.5 TRANSDERMAL DRUG DELIVERY

Topically applied medications, which are administered to the skin to deliver the medication at a determined rate and in a regular manner, are known as transdermal drug delivery systems (TDDS). Medications can be delivered through various routes such as orally, parenterally, nasal, transdermal, rectal, ocular, and intravaginal. Among these routes, the most common and popular one is the oral route. However, this route has some limitations, such as drug degradation before reaching the bloodstream and drug degradation in the gastrointestinal tract (GIT) due to acidic pH and various enzymatic reactions present in the gastrointestinal tract. To minimize these limitations, a novel drug delivery system (TDDS) has been developed, in which a polymer is combined with a drug in such a way that the drug is released from the polymer matrix in a regular manner.

10.5.1 BENEFIT OF TRANSDERMAL DRUG DELIVERY SYSTEM (TDDS)

- Transdermal delivery system avoids first-pass metabolism of drugs through gastrointestinal pH, various enzymatic reactions, and other oral administration of drugs.
- Avoid gastrointestinal incompatibility.
- Decreasing undesirable toxic effects or side effects.
- Acceptable for drugs that contain a very short half-life and narrow therapeutic range.

10.5.2 DISADVANTAGES OF TRANSDERMAL DRUG DELIVERY SYSTEM (TDDS)

- This system is not applicable for delivery of various types of ionic drugs.

- Only potent drugs, i.e., those that work at a low dose, are acceptable for transdermal delivery due to the skin's impermeability.

10.6 OTHER CONTROLLED DRUG DELIVERY SYSTEMS (DDS)

10.6.1 ORAL

Modified oral release drug delivery system (MODDS) designed to increase the release of drug for many hours. The modified release system offers a decrease in dosing frequency. Decrease risk of toxic and side effects and improve medicated effect in the body and increase bioavailability. The physiological factors which are in control absorption of drug include gastric and intestinal transit time. The rheological factors include solubility, stability, ionization, and lipophilicity. By controlling both physiological and rheological factors, design modified release of drug delivery system [8].

10.6.2 PARENTERAL CONTROLLED DRUG DELIVERY SYSTEM (PCDDS)

- The intravenous route (IV), subcutaneous route (SC), intramuscular (IM), intraperitoneal (IP), and intrathecal routes are some examples of parenteral routes of drug administration.
- Aqueous (water-soluble) and oil suspensions, oily solutions, and implants using this type of delivery system.
- The rate of absorption of various therapeutic activities will be determined.
- The parenteral administration, i.e., intravenous or IM route, is the most used and effective for delivery of pharmaceutical drug molecules which have low bioavailability and the drugs which have a narrow therapeutic ratio.

10.6.2.1 TYPES OF PARENTERAL CONTROLLED DRUG DELIVERY SYSTEM (PCDDS)

- Injectable drug delivery (drug which direct inject into skin);
- Implantable drug delivery system (drugs which apply onto skin) [9].

10.6.3 COLON-SPECIFIC DRUG DELIVERY

Targeted drug delivery to the colon is beneficial for curing various intestinal diseases such as ulcerative colitis, Crohn's disease, amoebiasis, colonic cancer, and systemic delivery of protein and peptide drugs.

The colon is the appropriate site for absorption of different peptide drugs and protein molecules for the given reasons: (i) less variation and potency of digestive enzymes; and (ii) enzymatic activity of colon mucosa is too less than that seen in the small intestine and its parts, thus CDDS protects protein drugs from various disintegrating and proteolytic degradation in the duodenum and jejunum parts of the intestine, and controls the release of the drug into the ileum or colon parts of the intestine, which increases systemic bioavailability. Drug preparation for intrarectal (IR) application is supplied in solutions, foam, and suppository forms. The concentration of the drug in the colon depends on formulation factors, the extent of drug expansion, and the holding time. Suppositories and foam drug molecules have been shown to be retained mainly in the rectum part of the intestine and sigmoid colon, while enema solutions have a great expanding capacity.

KEYWORDS

- **blood brain barrier**
- **dosage form**
- **drug delivery**
- **liposome**
- **nanoparticle**
- **parenteral controlled drug delivery system**
- **transdermal drug delivery system**

REFERENCES

1. Lia, C., & Wang, J., (2019). Recent progress in drug delivery annual review. *Acta Pharmaceutical Sinica B, 9*(6), 1145–1162.
2. Bhagwat, R. R., (2013). Novel drug delivery systems: An overview. *International J. Pharm. Sci Res., 4*(3), 970–982.

3. Verma, N. K., (2015). Liposomes: A targeted drug delivery system: A review. *Acta Medical Scientia, 2*(3), 65–70.
4. Patil, J. S., & Sarasija, S., (2012). Pulmonary drug delivery strategies: A concise, systematic review. *Lung India, 29*, 44–49.
5. Xiaowei, D., (2018). Current strategies for brain drug delivery. *Theranostics, 8*(6), 1481–1493.
6. Tiwari, G., & Tiwari, R., (2012). Drug delivery systems: An updated review. *International J. Pharma Investigation, 2*, 2–11.
7. Ghume, V. K., (2020). *American Journal of Pharmatechnology Research Review Article, 10*(2), 33–47.
8. Kalyani, M., (2013). Parenteral controlled drug delivery system. *International Journal of Research in Pharmaceutical and Nano Sciences, 2*(5), 572–580.
9. Philip, A. K., (2010). *Colon Targeted Drug Delivery Systems: A Review on Primary and Novel Approaches, 25*, 70–78.

CHAPTER 11

A REVIEW ON THIADIAZOLE CONTAINING COMPOUND AS POTENT ANTINEOPLASTIC AGENTS

DIVYA SAHU,[1] SAGAR SAHU,[2] PREETI PATEL,[3] VIJAY K. PATEL,[4] and HARISH RAJAK[5]

[1]*Rungta College of Pharmaceutical Science and Research, Raipur, Chhattisgarh, India*

[2]*Columbia Institute of Pharmacy, Raipur, Chhattisgarh, India*

[3]*ISF College of Pharmacy, Moga, Punjab, India*

[4]*Dr. C. V. Raman Institute of Pharmacy, Dr. C. V. Raman University, Kota, Bilaspur, Chhattisgarh, India*

[5]*Department of Pharmacy, Guru Ghasidas University, Bilaspur, Chhattisgarh, India*

ABSTRACT

Cancer remains one of the leading causes of death despite significant advances in understanding its pathophysiology. As a result, novel, more effective anticancer medicines are immediately required. The thiadiazole ring is a flexible scaffold and thiadiazole-containing compounds can permeate the cell membrane and interact with biological receptors because of the ring's mesoionic nature. As a result, these substances have a wide range of biological effects. This review describes the current scientific information about thiadiazole derivatives that demonstrate effectiveness *in vitro* and/or *in vivo* across cancer models, focusing on their targets of action. It is shown

Computational Optimization, Modeling, and Simulation for Engineering Applications.
Anupam Shukla, Sourabh Rungta, Mohan Awasthy, and Rakesh Himte (Eds.)

how the substituent affects the activity of the substances. Furthermore, the findings of clinical trials evaluating thiadiazole-containing cancer moieties.

11.1 INTRODUCTION

In 2018, 18.1 million new cancer cases and 9.6 million cancer deaths were reported within the International Agency for Research on Cancer (IARC). The world population is growing continuously, so the rate of cancer cases is predicted to rise. In recent years, researchers have discovered a number of possible specific treatment targets for various types of cancer and have made significant advances in understanding their molecular biology, but due to multilabel drug resistance and adverse effects of chemotherapy, more anticancer therapy techniques are urgently needed.

The thiadiazole molecule is a five-membered heterocyclic molecule with two nitrogen atoms and one sulfur and two carbon. The thiadiazole molecule has four isoforms found in nature, i.e., 1,2,3-thiadiazole, 1,2,4-thiadiazole, 1,2,5-thiadiazole, and 1,3,4-thiadiazole [1] (Figure 11.1). The bioisostere of pyrimidine and oxadiazole, thiadiazole, possesses a diverse range of pharmacological properties, such as antibacterial, antifungal, antiparasitic, anti-inflammatory, anticancer properties, and antiviral [2]. The mesoionic character of thiadiazoles increases the permeability through cellular membranes. The presence of the sulfur atom in thiadiazoles is most likely to blame for their high liposolubility [3]. Thiadiazole-containing medicines are currently used in clinics, including diuretics acetazolamide and methazolamide, as well as antibiotics cefazedone and cefazolin sodium. There have been several scientific studies that have demonstrated the potential anticancer properties of thiadiazole derivatives, while many are undergoing clinical trials as single agents or combined with other anticancer agents.

FIGURE 11.1 Schematic diagram of different isoforms of Thiadiazole.

11.2 DERIVATIVES OF 1,2,3-THIADIAZOLE

Microtubules, which make up the filamentous cytoskeleton, are formed by tubulin. There are many biological functions performed by microtubules, including intracellular polarity and cell shape. Different heterocyclic compounds blocking tubulin polymerization have anticancer potential, according to an increasing amount of research.

Wu and colleagues focused on analogs of combretastatin A-4 (CA-4), a tubulin-binding anticancer drug that inhibits microtubule polymerization. The CA-4 activity depends on the arrangement of the double bond between the olefin group and the 3,4,5-trimethoxyphenyl group. The olefin group of CA-4 is substituted with 1,2,3-thiadiazole group showed cytotoxicity in a variety of cell lines, including human myeloid leukemia HL-60, human colon adenocarcinoma HCT-116, and immortalized human microvascular endothelial (HMEC-1) cell lines. The 1,2,3-thiadiazole CA-4 derivatives compound 3e, which included a 4-$EtOC_6H_4$ group at position 5 of the thiadiazole ring and an ethyl substituent at position 5 of the dihydroxyphenyl ring have had cytotoxic activity similar to CA-4 or lower in all three cell lines (GI50 values for HeLa and U2OS cells were 0.70 M and 0.69 M, respectively).

The 5-aryl-4-(5-substituted-2,4-dihydroxyphenyl)-1,2,3-thiadiazoles with a chloro-substituent at position 3 of the dihydroxyphenyl moiety were a mild inhibitor of cancer cell viability. The presence of this chloro-substituent most likely inhibited the creation of a large H-bonding network, resulting in a lack of activity [4]. Three of the compounds that were shown to bind Hsp90 the most effectively were studied further. They displayed antiproliferative activity against human colon cancer HCT-116 cells with GI50 values ranging from 3.2 to 4.6 μM. The mechanism of antiproliferative action of the compounds was confirmed by the reduction of Hsp90 client proteins, CRAF, ERBB2, and CDK4, in HCT-116 cells. Furthermore, higher expression of Hsp27 and Hsp72 was seen after such treatment, indicating that the heat shock response was activated. In addition, the induction of apoptosis was demonstrated by an increase in PARP cleavage elicited by the examined substances cytotoxic activity IC_{50} ranging from 13.4 to 86.6 nM[5]. The 3,4,5-trimethoxyphenyl moiety substituted with 1,2,3-thiadiazole at 4th position showed substantial cytotoxic activity cancer cell lines, while the substitution at 5th position of 4,5-trimethoxyphenyl moiety having less or not showing cytotoxity. These chemicals inhibited tubulin polymerization with activity comparable to CA-4 and stopped the cell cycle at the G2/M phase. Two of these compounds inhibited tumor development in the S180 sarcoma model in mice at a rate equivalent to or even greater than CA-4 [6].

11.3 INHIBITORS OF HSP90

The 1,2,3-thiadiazole synthetic analogues and derivatives inhibit the activity of heat shock protein 90 (Hsp90). Hsp90 is a chaperone that regulates how many proteins fold. Several oncoproteins are degraded when its activity is inhibited. Cancer cells are more sensitive to Hsp90 inhibition than normal ones, suggesting that this protein might be a viable anticancer target [7].

11.4 DERIVATIVES OF 1,2,4-THIADIAZOLE

Resveratrol is a trans stilbene derivative found in nature that has anti-cancer and cancer chemo-preventive properties. The bioavailability of stilbene derivative is low due to quick and widespread metabolism. An increasing amount of information suggests that resveratrol's bioavailability and efficacy against many cancers can be considerably improved by various chemical changes. Mayhoub & colleagues replaced the trans stilbene ethylenic bridge in the resveratrol scaffold with a 1,2,4-thiadiazole heterocycle, adjusted the substituents on the two aromatic rings, and assessed the compounds' cytotoxicity against the MCF-7 human breast cancer cell line. The 3-hydroxy-derivative of resveratrol have potential anticancer activity as IC_{50} values ranging from 4.7 to 39.8 μM. The structural changes. Benzo[4,5] derivatives with three substitutes imidazo[1,2-d] [1,2,4] of resveratrol showed potential cytotoxic activity. The 1,2,4-thiadiazole moiety is thiadiazole coupled to bromoacryloyl amido benzoheterocycles with a polymethylene spacer showed potential cytotoxic activity in the human myeloid leukemia cell lines HL-60 and U937 and SK-MEL-1. These compounds 1 and 2, which include a benzo-heterocycle with N-unsubstituted indole, have potential cytotoxic activity. However, when the chain length of compounds 1 and 2 was increased from three to six methylene units, the anticancer activity of these two derivatives reduced. The compound 2 containing the bromoacrylamidoindole moiety had the highest potency (IC_{50} values ranging from 0.40 to 3.02 M) among the derivatives defined by six methylene units [8]. The antitumor activity was reduced when indole was substituted in compound 2 with N-methyl indole (compound 3). When indole was substituted with benzofuran (compound 4) and benzothiophene, the reduction was even greater (compound 5). The compound 1 appeared to be a less efficient apoptosis inducer than the other derivatives for treatment of HL-60 and U937 cells suggesting that this molecule may stimulate alternative pathways causing cell death [9].

11.5 DERIVATIVES OF 1,2,5-THIADIAZOLE

The compound NSC745885 is anthra[1,2-d] imidazole-6,11-dione tetracyclic analogs with varied side chains, thiadiazole-6,11-dione, having cytotoxic potential in the treatment of leukemia, melanoma, ovarian, breast, prostate, and glioma cell lines. The sensitivity to NSC745885 in colon and non-small cell lung cancer (NSCLC) cell lines activity is GI50 values between 1.28 and 17.40 M) [10]. The NSC745885 compound was successful in oral cancer cell lines but had no effect on normal lung fibroblast survival. *In vivo*, the NSC745885 therapy reduced tumor growth but had no effect on the animals' body weight. This substance possesses anti-tumor activity similar to that of doxorubicin, an anthraquinone derivative, but with a higher level of safety. The treatment with NSC745885 resulted in the induction of apoptosis both *in vitro* and *in vivo*[11]. In another study from the same group, the chemical NSC745885 reduced the viability of bladder cancer T24 and MBT2 cell lines and caused G2/M cell cycle arrest but had no effect on immortalized normal urothelial cells SV-HUC-1 or normal fibroblasts 3T3. The same detrimental impact on vitality was observed when cancer cells were treated with 2.5 M doxorubicin or 40 M emodin, a natural anthraquinone, however both of these compounds induced significant toxicity in SV-HUC-1 and 3T3 cells.

11.6 INHIBITORS OF INOSINE MONOPHOSPHATE DEHYDROGENASE (IMPDH)

Shapiro published one of the first studies on 1,3,4-thiadiazole derivatives' anticancer properties. The niacin analog 2-ethylamino-1,3,4-thiadiazole (EATDA) was shown to protect mice from developing mammary adenocarcinomas in this study. The antitumor activity of EATDA was preceded by nicotinamide therapy, indicating that it is a niacin antagonist. The EATDA was also found to improve the antitumor activity with the combination of 8-azaguanine or deoxypyridoxine or testosterone [12]. The EATDA and 2-amino-1,3,4-thiadiazole (ATDA) were shown to have anti-cancer effects in mice with the systemic leukemia L1210. These compounds not only helped the animals live longer but also blocked tumors from forming. The treatment of nicotinamide prevented the anti-leukemic action and host toxicity of both compounds [13]. Treatment of L1210-bearing animals with ATDA, also known as NSC 4728, reduced adenine and guanine ribonucleotide pools in tumor cells while increasing uridine triphosphate (UTP) and inosine monophosphate (IMP) pools. Simultaneous administration of nicotinamide was shown to counteract this effect. Given that nicotinamide inhibits ATDA's anti-leukemic activity, it's possible that ATDA's anti-leukemic activity is linked to guanosine monophosphate production. The compound NSC 4728 and its analogs appeared to be strong inhibitors of IMP dehydrogenase [14, 15] (Figure 11.2).

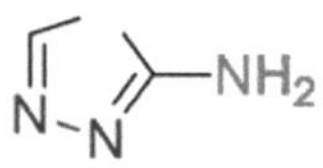

FIGURE 11.2 Structure of 2-amino-1,3,4-thiadiazole (ATDA: NSC 4728).

11.7 CONCLUSIONS

A wide range of thiadiazole compounds exhibit anticancer effects. However, the research described above has a number of limitations. First, caution is suggested when comparing the efficacy of medications as determined by various test methodologies. Second, the majority of thiadiazole derivatives have been studied in cancer cells, but their effects in non-cancerous cells are unknown. Third, in the bulk of the studies cited, the medications' efficacy was determined *in vitro*. Finally, the majority of thiadiazole derivatives' molecular structures are yet to be determined. To assess the selectivity, bioavailability, and mechanism of action of drugs, further research is required. Despite this, the findings show that the thiadiazole moiety might be a useful lead structure in the creation of novel anticancer drugs.

KEYWORDS

- **inosine monophosphate dehydrogenase**
- **potent antineoplastic agents**
- **ribonucleotide pools**
- **thiadiazole**
- **tumor cells**
- **uridine triphosphate**

REFERENCES

1. Bray, F., Ferlay, J., Soerjomataram, I., Siegel, R. L., Torre, L. A., & Jemal, A., (2018). Cancer statistics 2018: GLOBOCAN estimates of incidence and mortality worldwide for 36 cancers in 185 countries. *CA Cancer J. Clin., 68*(6), 394–424.
2. Li, Y., Geng, J., Liu, Y., Yu, S., & Zhao, G., (2013). Thiadiazole—A promising structure in medicinal chemistry. *ChemMedChem., 8*(1), 27–41.
3. Haider, S., Alam, M. S., & Hamid, H., (2015). 1, 3, 4-thiadiazoles: A potent multi targeted pharmacological scaffold. *European Journal of Medicinal Chemistry, 92*, 156–177.
4. Haider, K., Rahaman, S., Yar, M. S., & Kamal, A., (2019). Tubulin inhibitors as novel anticancer agents: An overview on patents (2013–2018). *Expert Opinion on Therapeutic Patents, 29*(8), 623–641.
5. Wu, M., Sun, Q., Yang, C., Chen, D., Ding, J., Chen, Y., Lin, L., & Xie, Y., (2007). Synthesis and activity of combretastatin A-4 analogues: 1, 2, 3-thiadiazoles as potent antitumor agents. *Bioorganic & Medicinal Chemistry Letters, 17*(4), 869–873.
6. Bagatell, R., & Whitesell, L., (2004). Altered Hsp90 function in cancer: A unique therapeutic opportunity. *Molecular Cancer Therapeutics, 3*(8), 1021–1030.
7. Cikotiene, I., Kazlauskas, E., Matuliene, J., Michailoviene, V., Torresan, J., Jachno, J., & Matulis, D., (2009). 5-Aryl-4-(5-substituted-2, 4-dihydroxyphenyl)-1, 2, 3-thiadiazoles as inhibitors of Hsp90 chaperone. *Bioorganic & Medicinal Chemistry Letters, 19*(4), 1089–1092.
8. Sharp, S. Y., Roe, S. M., Kazlauskas, E., Čikotienė, I., Workman, P., Matulis, D., & Prodromou C. (2012). *Co-crystalization and in Vitro Biological Characterization of 5-aryl-4-(5-Substituted-2-4- Dihydroxyphenyl)-1, 2, 3-Thiadiazole Hsp90 Inhibitors, 7*(9), 1–8.
9. Dai, H., Ge, S., Li, G., Chen, J., Shi, Y., Ye, L., & Ling, Y., (2016). Synthesis and bioactivities of novel pyrazole oxime derivatives containing a 1,2,3-thiadiazole moiety. *Bioorganic & Medicinal Chemistry Letters, 26*(18), 4504–4507.
10. Ferraz Da, C. D. C., Pereira, R. L., Martins-Dinis, M. M., Ferretti, G. D., Ferreira, V. F., & Silva, J. L., (2020). Anticancer potential of resveratrol, β-lapachone and their analogues. *Molecules, 25*(4), 893.
11. Mayhoub, A. S., Marler, L., Kondratyuk, T. P., Park, E. J., Pezzuto, J. M., & Cushman, M., (2012). Optimizing thiadiazole analogues of resveratrol versus three chemopreventive targets. *Bioorganic & Medicinal Chemistry, 20*(1), 510–520.
12. Romagnoli, R., Baraldi, P. G., Carrion, M. D., Cruz-Lopez, O., Preti, D., Tabrizi, M. A., Fruttarolo, F., et al., (2007). Hybrid molecules containing benzo [4, 5] imidazo [1, 2-d] [1, 2, 4] thiadiazole and α-bromoacryloyl moieties as potent apoptosis inducers on human myeloid leukaemia cells. *Bioorganic & Medicinal Chemistry Letters, 17*(10), 2844–2848.
13. Lee, Y. R., Chen, T. C., Lee, C. C., Chen, C. L., Ali, A. A., Tikhomirov, A., Guh, J. H., et al., (2015). Ring fusion strategy for synthesis and lead optimization of sulfur-substituted anthra [1, 2-c] [1, 2, 5] thiadiazole-6, 11-dione derivatives as promising scaffold of antitumor agents. *European Journal of Medicinal Chemistry, 102*, 661–676.

14. Lin, P. L., Chang, J. T., Wu, D. W., Huang, C. C., & Lee, H., (2016). Cytoplasmic localization of Nrf2 promotes colorectal cancer with more aggressive tumors via upregulation of PSMD4. *Free Radical Biology and Medicine, 95*, 121–132.
15. Gelibter, A. J., Caponnetto, S., Urbano, F., Emiliani, A., Scagnoli, S., Sirgiovanni, G., Napoli, V. M., & Cortesi, E., (2019). Adjuvant chemotherapy in resected colon cancer: When, how and how long? *Surgical Oncology, 30*, 100–107.

CHAPTER 12

DESIGN AND ANALYSIS OF MUFFLER FOR EXHAUST SYSTEM

NIKHIL NAGDEV, TEJAS PATIL, JAYESH DEORE, SHUBHAM BHOJANE, TANMAY BARGE, and AVINASH SOMATKAR

Department of Mechanical Engineering, Vishwakarma Institute of Information Technology, Pune, Maharashtra, India

ABSTRACT

Today's world needs some differentiation even in exhaust system noise and also needs to meet the strict noise standards of the government and the cost of the system. In an automobile, the muffler plays a major role in minimizing the automobile sound and creating a pleasant signature noise to make the rider comfortable. To satisfy all of these requirements, our muffler design comes into play. This chapter describes how to design, develop, and validate a muffler for a specific type of engine. The chapter also discusses various operational requirements for mufflers, such as material, weight, insertion loss, backpressure, and size. In this design, we are using a hybrid muffler, which combines absorptive and reactive types of mufflers. We chose a hybrid muffler because it can provide better sound attenuation than a reactive or absorptive muffler alone. Several designs were considered, and the best design was selected. Its CAD model was created using SolidWorks. After that, all virtual simulations were performed using ANSYS software. Considering economic conditions, stainless steel material 302 was selected for the entire body of the system, and glass fiber was used as the absorptive material. In the end, detailed results are obtained and discussed, demonstrating the attenuation performance of the exhaust system.

Computational Optimization, Modeling, and Simulation for Engineering Applications.
Anupam Shukla, Sourabh Rungta, Mohan Awasthy, and Rakesh Himte (Eds.)

12.1 INTRODUCTION

Noise is an annoyance since the invention of the automobile till present days over the streets and the internal combustion engine is the main contributor to it. Although, the noise due to exhaust gasses is ten times of all the other contiguous noises coming from vehicle, so there may be the possibility of hearing disorder if anyone is exposed to excessive noise above a particular level. Hence the proper design, validation and analysis of the automotive muffler is must. Along with sound attenuation, there is one more important function of the muffler to produce optimum back pressure as it effects the performance of the internal combustion engine. Hence the good design of muffler is must to achieve all suck kinds of needs. So in this paper, we aim to study and examine the design of mufflers with various geometries. Also to study the geometrical effect of the muffler on their noise transmission loss maximization performance [1, 2, 5, 8].

12.2 NEED OF MUFFLER

Muffler is one of the most important parts of the automobile and is often used in the exhaust system to reduce noise caused due to exhaust gases. The constructional design of the muffler is very important as it affects the efficiency of the engine. Noise characteristics such as emission and fuel mileage of the engine are also affected by design consideration of muffler. Engine sound muffles with intricate metal system and tubes allow the engine sound waves to deviate from all obstruction. Secondly, the muffler maintains the engine's backpressure. Backpressure simply means the difference between the atmospheric pressure and pressure inside the muffler, it is created in the opposite direction of the fluid flow when there are restrictions to the fluid flow, while moving at high velocity. It also helps to amplify and tune the frequencies of exhaust waves and makes a smooth sound from it. Therefore, to achieve all these advantages a good design of muffler is needed. Conventionally, muffler design and validation has been the repetitive process which cost a lot of human power time and money. This design and analysis paper will help us to reduce the noise levels in upcoming vehicles by an easy and simple process [8, 9].

12.3 TYPES OF MUFFLERS

There are various types of mufflers available in the market. Here we have discussed three main types of mufflers, which are as follows in subsections.

12.3.1 REACTIVE MUFFLER

The reactive muffler is also known as the reflective muffler. In the reactive muffler, in the inner section of the casing of the muffler, various shaped chambers are provided. This type of muffler uses the principle of destruction. Due to this destructive phenomenon of pressure, waves cancel themselves and result in noise drop. This type of muffler simply uses the resonating effect to reduce noise by providing a resonating area, hence it is also called a resonance muffler.

12.3.2 ABSORPTION MUFFLER

As its name suggests, the absorption of the pressure wave is used to reduce exhaust gas noise. In the case of an absorption-type muffler, a straight perforated pipe is used which directly connects the inlet and outlet exhaust pipes. Around this pipe, noise-absorbing material is used which absorbs the noise from the exhaust pipe. Commonly used materials for noise absorption are glass wool, rock wool, etc. Simply put, noise is reduced by converting energy into heat energy by employing absorbing material. In an absorption muffler, there are no restrictions to gas flow due to the straight pipe, hence this muffler doesn't increase the back pressure and also improves vehicle performance. However, the problem with this muffler is that it reduces less noise compared to other mufflers.

12.3.3 COMBINATION MUFFLER

As the name suggests, this muffler combines the properties of both reactive and absorptive types of mufflers. Reactive mufflers cause a very large generation of back pressure, which reduces vehicle performance, while the absorption type of muffler does not properly attenuate the noise and can only run at low frequencies. So, the combination muffler combines the good properties of both mufflers and removes all the disadvantages, resulting in the best possible results.

12.4 FUNCTIONAL REQUIREMENTS OF MUFFLER

12.4.1 ADEQUATE INSERTION LOSS (ATTENUATION ABILITY)

The role of a muffler in an exhaust system is to muffle or attenuate the noise. The reduction in sound level from the noise source can only be achieved by a good exhaust application. Hence, the muffler must be capable of reducing the noise within the system produced by the engine. Then only it can satisfy the government standard insertion losses demand. Insertion loss simply means the capability of the muffler to attenuate the sound. Insertion loss can also be defined as the difference (in decibels) between the noise before entrance of the muffler and leaving the muffler.

12.4.2 BACK PRESSURE

In simple terms, back pressure can be elaborated as the difference between the atmospheric pressure and the pressure inside the muffler, and it occurs due to holes in the pipe in the muffler and sudden discontinuities in space. The higher the back pressure, the more noise reduction takes place. On the other hand, back pressure is directly proportional to engine performance. An increase in back pressure decreases engine efficiency and vice versa.

12.4.3 DURABILITY

The most important thing in any kind of device is its reliability, how much the device will run affects the selection criteria. Therefore, durability plays a very vital role, especially in the case of a muffler where the temperature of exhaust gases inside the muffler is very high. Material selection also affects the durability of the muffler. So proper material should be selected while considering the design of the muffler.

12.4.4 COST

We know performance is the important factor, but also the cost of a product is a major factor to be considered. The cost of the muffler depends upon the complexity of design, materials selection, manufacturing process, muffler reliability, and labor costing.

12.5 MUFFLER DESIGN METHODOLOGY

The Muffler Design methodology contains a total of 7 steps to reach the final product of a muffler. These steps are discussed broadly in subsections to get the best possible design of a muffler.

12.5.1 STEP 1: BENCHMARKING

Benchmarking is the core step in any design process, its aim is to set the targets by doing benchmarking exercises of the similar kind of models. It is one of the most important components for improvement of the programs, it is a key to satisfy the quality assurance.

This step is designed to set a goal in comparison to standard parameters. Based on the engine input data provided and benchmark research, the following points are benchmarked:

- Transmission loss;
- Back pressure;
- Output noise range in dB.

12.5.2 STEP 2: CALCULATE TARGET FREQUENCIES

After the benchmarking exercise, the second step is to calculate the target frequencies. Target frequencies are calculated to understand how much transmission loss is occurring in the system. Target frequencies are calculated using the following formulae:

1. **Cylinder Firing Rate:**

$$\text{C.F.R} \frac{\text{Engine speed in RPM}}{60} \quad \text{(for a 2-stroke engine)}$$

$$\text{C.F.R} \frac{\text{Engine speed in RPM}}{120} \quad \text{(for a 4-stroke engine)}$$

2. **Engine Firing Rate:**

$$\text{E. F. R} = n * (\text{C. F. R}) \quad (n = \text{number of cylinders})$$

12.5.3 STEP 3: MUFFLER VOLUME CALCULATION

Here are all the formulas that were used to find the volume of muffler. These calculations are based on the past experiences and acoustics theory of muffler design for various engines.

1. **Swept Volume (Vs):**

$$V_s = \frac{\pi \times D_2 \times L}{4} \quad \text{(D: Engine bore diameter; and L: stroke length).}$$

2. **Volume to be Used for Calculation (V):**

$$V = n \times \frac{V_s}{2}$$

3. **Muffler Volume (Vm):**

$$V_m = V \times (12 \text{ to } 25)$$

(Actual volume is increased by 12 to 25 times by the muffler volume depending on space availability. It is because of the safety and other allowances.)

4. **Diameter of the Muffler (d):**

$$V_m = \frac{\pi \times D_2 \times 1}{4}$$ (d: diameter of muffler; and l: length of muffler)

12.5.4 STEP 4: INTERNAL DESIGN OF MUFFLER

The next step is internal configuration, this step is performed to obtain the internal design of muffler, and it is done to obtain a good internal configuration so that the best acoustic performance can be achieved.

1. **Hole Diameter on the Perforated Pipe is Calculated by:**

$$d_1 = \frac{316}{\sqrt{N}}$$

2. **Porosity (σ) is Given by:**

$$\sigma = \frac{\pi (4 \times d_1)^2}{c^2}$$

As we know, the main cause of backpressure is the restriction, so to achieve the optimum backpressure, the porosity needs to be maintained. In general, a large porosity is preferred.

3. **AOP (Area of Preformation) is Given by:**

$$\text{A.O.P} = \frac{\text{Area of performated holes}}{\text{Plain area without holes}}$$

The AOP is inversely proportional to the transmission loss. So minimum the AOP better will be transmission loss and subsequently better the acoustic performance. Hence designer needs to manage both the properties during design of muffler.

12.5.5 STEP 5: VIRTUAL SIMULATION

After completing all the above processes and calculations, different design concepts are derived with various elements inside the volume of the muffler. Then, CAD models are designed for these concepts, and these models are virtually simulated using CFD software such as ANSYS. The following analyses were conducted.

1. **CFD Analysis:** In the CFD analysis, the properties like back pressure and velocity are analyzed at different positions of the muffler and checked what differences are occurring in the process.
2. **Transmission Loss Analysis:** Transmission loss plays a vital role as it affects the engine efficiency and also to satisfy the government norms, so we must need to predict how much the transmission loss will occur. The number of software's are accessible in the market for forecasting the transmission loss, designers can select the proper software as per their need.

12.5.6 STEP 6: PROTOTYPE MANUFACTURING

After finalizing the best design model for muffler, the next step comes as prototype manufacturing. A prototype is a small model which is built to see the effects in the real world.

12.5.7 STEP 7: EXPERIMENTAL TESTING

This is the final step, in which the final experimental testing is done on the prototype model. In experiments, the back pressure and transmission loss of all concepts (CAD models) that are made at the above step are checked. If all the results are as per the required standards, then finally the design is selected and can be used for a particular type of engine and can be used for mass production.

12.6 CASE STUDY: LCV DIESEL ENGINE VEHICLE

For understanding the design methodology, we have selected a case study of Hyundai Creta car and based on the case study, we have designed our model (Table 12.1).

TABLE 12.1 Engine Specifications

Bore diameter (D)	69.6 mm
Stroke length (L)	82 mm
No of cylinders (n)	4
Engine power (P)	88.7 HP
Max speed (N)	4,000 rpm

12.6.1 STEP 1: BENCHMARKING

As per the above-discussed design methodology, we standardized our muffler model compared with the same kind of models available in the market to set the target of insertion loss of the muffler.

Max back pressure for muffler = 0.0248 bar.

Insertion loss (Noise reduction target for muffler) = 30 dB.

12.6.2 STEP 2: CALCULATE TARGET FREQUENCIES

1. **Cylinder Firing Rate:**

$$\text{C.F.R}\frac{\text{Engine speed in RPM}}{120}=\frac{4000}{120}=33.33\,\text{H}_Z$$

2. **Engine Firing Rate:**

$$\text{E.F.R=n}\times(\text{C.F.R})=4\times33.33=133.33\,\text{H}_Z$$

12.6.3 STEP 3: MUFFLER VOLUME CALCULATION

1. **Swept Volume (Vs):**

$$\text{V}_\text{s}=\frac{\pi\times\text{D}^2\times\text{L}}{4}=\frac{3.14\times69.6^2\times82}{4}=0.3118\,\text{liters}$$

2. **Total Swept Volume in Liters:**

$$V_{ts} = 4 \times V_s = 1.2472 \text{ liters}$$

3. **Volume to be Consider for Calculation (V):**

$$V = n \times \frac{V_S}{2} 4 \times \frac{0.3118}{2} = 0.6236 \text{ liters}$$

4. **Muffler Volume (Vm):**

$$V_m = V \times (22) = 0.6236 \times 22 = 13.7192 \text{ (Factor considered is 22)}$$

5. **Diameter of the Muffler (d):**

$$d = \frac{\sqrt{V_m \times 4}}{\pi \times 1} = \frac{\sqrt{0.0137 \times 4}}{3.14 * 0.5} = 0.186\,\text{m} = 186\,\text{mm}$$

12.6.4 STEP 4: INTERNAL DESIGN OF MUFFLER

1. **Hole Diameter on the Perforated Pipe is Calculated by:**

$$d_1 = \frac{316}{\sqrt{N}} = \frac{316}{\sqrt{4000}} = 4.99 \sim 5\text{mm}$$

12.6.5 STEP 5: VIRTUAL SIMULATION

1. **CAD Model:** After doing calculations, the CAD model is built using SolidWorks, and here are all the design dimensions (Figure 12.1):

   Muffler:

   - Overall length = 600 mm;
   - Overall width = 258 mm;
   - Inlet pipe OD = 62 mm;
   - Outlet pipe OD = 62 mm;
   - Wall thickness = 5 mm.

   Perforated Pipe:

   - Length = 450 mm;

- OD = 62 mm;
- ID = 58 mm;
- Thickness = 2 mm;
- Perforated hole diameter = 5 mm;
- Distance between two holes = 10 mm.

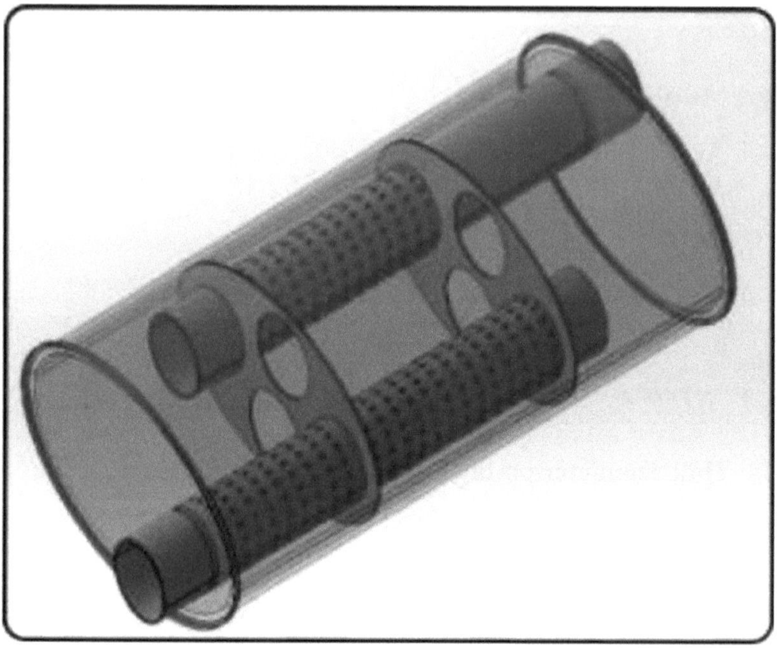

FIGURE 12.1 CAD model of muffler.

2. **Material Selection:** Mainly two materials are used in the whole design of the exhaust system, which are described below:

   i. **Stainless Steel 302:** It is a metal alloy formed of Iron and Chromium. It contains approximately 18% of Chromium and 8% of Nickel along with some High Carbon. Besides, it also contains manganese, silicon, sulfur, phosphorus, etc., in very small amounts. When stainless steel is heated, the chromium present in stainless steel produces a layer of chromium oxide outside the muffler, acting as a protective layer that delays its further oxidation. This layer also protects the metal from corrosion. The main highlight of grade 302 stainless steel is its high resistance to solvents, acids, and chemicals. Hence, it can be used for the manufacturing of the exhaust system rather than mild steel (Table 12.2).

TABLE 12.2 Stainless Steel Properties

Density	**7900.00 Kg/m3**
Specific heat	500 J/(KgK)
Thermal conductivity	16.30 W/(mK)
Resistivity	7.20e–07 Ohm.m

ii. **Glass Fiber:** It is used as sound absorption material. Since a purely reactive muffler cannot attenuate most of the sound by itself, we add a sound absorbing material inside the muffler which makes the muffler a hybrid muffler and it attenuates the sound to a great extent. Since the temperature in the muffler is high, we need to use a sound absorbing material which is resistant to heat and won't burn inside the muffler. Along with this, we don't want to increase the back pressure in the muffler because even a small increase in back pressure can affect the vehicle performance (Table 12.3).

TABLE 12.3 Glass Fiber Properties

Density	**1,281 to 1,762 kg/m3**
Withstanding temperature	520?C
Flow resistivity	4,500 Rayl/m

12.7 ANALYSIS

In order to check how efficiently the designed muffler is working, virtual analysis needs to be carried out. For analysis purposes, the most widely used software named 'ANSYS 2021 R2' is being used. The major criteria for analysis are velocity and pressure analysis. Basically, some input parameters (pressure and velocity) depending on the requirement are provided at the inlet of the muffler and the results will be analyzed accordingly. As mentioned above, the inlet boundary conditions are velocity distribution and pressure distribution. The initial values for velocity at the inlet and pressure at the inlet as the boundary conditions at the inlet are shown in Table 12.4.

TABLE 12.4 Exhaust Gases Inlet Conditions

Temperature (K)	**470**
Viscosity (kg/ms)	2.70e-05
Enthalpy (j/kg)	749575.3
Density (kg/m^3)	0.696
Length (mm)	600
Velocity (m/s)	80
Pressure (bar)	3.2

The final CFD analysis is done in 'ANSYS Fluent' which is more user-friendly to analyze the pressure and velocity and less time-consuming. In order to get accurate results, all 100 iterations are carried out for the designed muffler.

12.7.1 VELOCITY DISTRIBUTION

As shown in Figure 12.2, velocity analysis, the exhaust gases initially at a velocity of (80 m/s) and temperature (470 K) pass through the inlet pipe which is a perforated one. Since the perforated hole diameter is very small (i.e., 20 mm) as compared to pipe diameter, it proves less effective for velocity changes. Still, a slight increase in velocity can be seen due to the drilled holes. After the inlet pipe, exhaust gases expand in the expansion chamber leading to a decrease in velocity.

Then the gas velocity increases up to 270 m/s as it passes through the outlet exhaust pipe due to its narrow section. Baffles provided inside the expansion chamber ensure the smooth flow of exhaust gases and prevent excess pressure generation inside the chamber (Figure 12.2).

12.7.2 PRESSURE DISTRIBUTION

Decreasing the exhaust gas pressure is an important task to carry out as it creates unnecessary load on the system or engine. The designed model of the muffler proves very efficient for this job. As shown in Figure 12.3, in pressure analysis, the exhaust gases are initially at a pressure of 360,000 Pa and temperature 470 K passes through the inlet perforated pipe. As the gas passes through the perforated pipe pressure starts decreasing due to the narrow section of the pipe also the drilled holes contribute to this decrease

FIGURE 12.2 Velocity distribution.

in pressure. After the inlet pipe, gas expands in the expansion chamber and a small increase in pressure can be observed. But soon the pressure drops again due to baffles provided in the chamber. Further pressure drops up to a minimum 103,000 Pa due to the narrow outlet section of the outlet pipe (Figure 12.3).

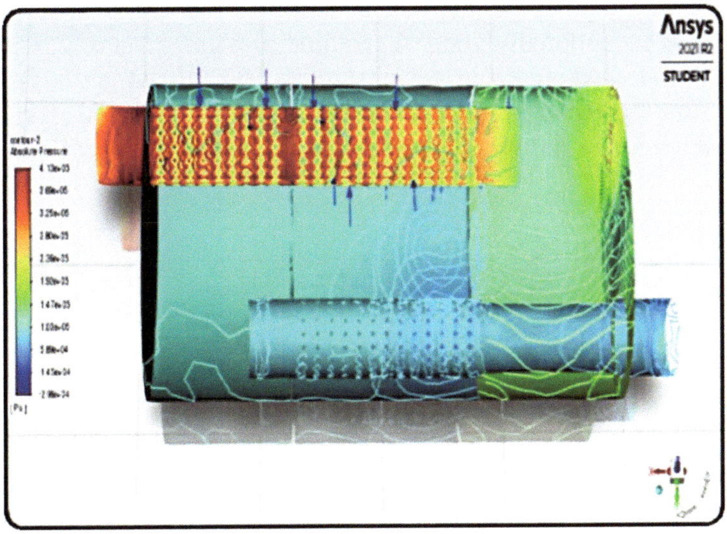

FIGURE 12.3 Pressure distribution.

12.8 WORKING OF OUR MODEL

As shown in the schematic diagram, the hybrid muffler consists of an expansion chamber, perforated tubes and baffles, and absorptive materials that are used to enhance the acoustic performance. The exhaust gases coming out from the engine first pass to the catalytic converter and then enter into the muffler. It enters into the muffler from the left side perforated pipe and then passes over to the last expansion chamber.

During passing through perforated pipe some waves exit from the pipe and got collide with walls causing friction, which destroys the sound waves. The expansion chamber plays the main crucial part in sound attenuation. The sound waves with more intensity reflect in the expansion chamber and cancel each other due to destructive interference. The final sound waves can then be absorbed by absorptive material when the exhaust gases pass through the final pipe.

12.9 CONCLUSION

In the conclusion, we can say that the properly designed muffler can overcome the present difficulties like more back pressure, high cost, and high transmission loss. By comparing, the results we obtained, the designed muffler is effective for engineering application. We also came to know that a hybrid muffler is far better than a reactive or absorptive muffler when working individually. Also, the cost required to design an individual absorptive and reactive will be higher as compared to the hybrid (absorptive + reactive) muffler. The results obtained from the analysis of our designed muffler are exceptionally good. A required pressure drop of 2,57,000 Pa is achieved and an increase in velocity up to 190 m/s which is satisfactory. Overall, the designed muffler proves excellent in virtual analysis, but the practical approach and physical model can affect the design considerations.

KEYWORDS

- **CFD analysis**
- **Muffler design**
- **insertion loss**
- **Muffler**
- **prototype manufacturing**
- **signature noise**
- **Muffler validation**

REFERENCES

1. Shinde, P. V., Gavali, P. M., Barawade, R. A., Mohite, Y. B., & Shinde, P. B., (2017). A review on muffler design for exhaust noise attenuation. *International Journal of Engineering and Technology, 9,* 428–431.
2. Potente, D., (2005). General design principles for an automotive muffler. In: *Proceedings of ACOUSTICS* (pp. 153–158).
3. Kumar, N. L., & Veladri, K., (2018). Modification of muffler design to increase exit velocity. In: *IOP Conference Series Materials Science and Engineering* (Vol. 455, No. 1, p. 012109). IOP Publishing.
4. Prajapati, V. D., & Desai, A. J., (2016). Design and analysis of automotive muffler. *Int. J. Eng. Res. Technol., 5.*
5. Tutunea, D., Calbureanu, M. X., & Lungu, M., (2013). The computational fluid dynamics (CFD) study of fluid dynamics performances of a resistance muffler. *International Journal of Mechanics*, (4).
6. Singh, J., Nigam, S. P., & Bhagi, L. K., (2017). *A Study on Effectiveness of Muffler on a Two-Wheeler Vehicle Noise, 4*(3), 24–29.
7. Anekar, N., & Nimbalkar, S., (2015). Vibration and noise in reactive muffler A study. In: *AMET Conference* (pp. 1–5).
8. Rajadurai, S., Raj, R. G., & Rajakumar, K. J. (2015). *Countermeasures to Reduce Muffler Shell Radiation Noise, 2*(7), 489–496.
9. Mohiuddin, A., Ideres, M., & Hashim, S., (2005). Experimental study of noise and back pressure for silencer design characteristics. *Journal of Applied Sciences (Pakistan), 5*(7), 1292–1298.
10. Zeng, F., & Sun, S., (2018). Study on the sound quality of steady and unsteady exhaust noise. *Mathematical Problems in Engineering, 2018.*
11. Suryawanshi, R., Kashikar, A., & Sunnapwar, V., (2020). *Design and Analysis of Automotive Mufflers for Noise Attenuation A Review.* Available at: SSRN 3707962.
12. Chaudhri, M. J. H., Patel, B. S., & Shah, S. A. (2014). *Muffler Design for Automotive Exhaust Noise Attenuation-A, 4*(1)(Version 2), 220–223.
13. Shah, S., Kuppili, S., Hatti, K., & Thombare, D., (2010). *A Practical Approach Towards Muffler Design, Development and Prototype Validation (No. 2010-32-0021).* SAE Technical Paper.
14. Sherekar, V., & Dhamangaonkar, P. R., (2014). Design principles for an automotive muffler. *International Journal of Applied Engineering Research, 9*(4), 483–489.
15. https//www.actonservicecentre.co.uk/ (accessed on 21 December 2023).

CHAPTER 13

IN-SILICO DESIGN OF FLUTAMIDE ANALOGUES USING BIOISOSTERIC APPROACH FOR PROSTATE CANCER THERAPY

AJAY KUMAR GUPTA and SANMATI K. JAIN

Department of Pharmacy, Guru Ghasidas Vishwavidyalaya (A Central University), Bilaspur, Chhattisgarh, India

ABSTRACT

Flutamide is a non-steroidal anti-androgen drug used to treat prostate cancer (PC). It causes hepatotoxicity and/or liver failure, which are the major causes of drug withdrawal. In the present study, newer analogs of Flutamide were generated using a bioisosteric approach. Around 77 bioisosteres of flutamide replacing the isopropyl group were generated using MolOpt. Pharmacokinetic (Absorption, Distribution, Metabolism, and Excretion) and toxicity properties were calculated for the newly designed analogs. Drug-likeness and drug score (DS) were also calculated. Results indicate that analogs 007 [N-(4-nitro-3-(trifluoromethyl) phenyl) cyclobutanecarboxamide] and 020 [2-ethyl-N-(4-nitro-3-(trifluoromethyl) phenyl) butanamide] showed good androgen receptor (AR) interaction (NR-AR > 0.7), less drug-induced liver toxicity (DILI), and QED value (≥ 0.67) comparable with Flutamide. Also, analogs such as 055, 076, and 037 were significantly active and less toxic than the standard drug. Further studies may lead to the development of newer non-steroidal anti-androgen analogs for prostate cancer.

Computational Optimization, Modeling, and Simulation for Engineering Applications.
Anupam Shukla, Sourabh Rungta, Mohan Awasthy, and Rakesh Himte (Eds.)

13.1 INTRODUCTION

In 2020, India had a population of 1,380 million, with 1.3422 million new cancer cases and 0.8516 million cancer deaths. Prostate cancer (PC) new cases increased by 34,540 (2.6%), while deaths increased by 16,783 (2.0%) [1]. The steroidal hormones testosterone and androgen promote the proliferation of prostate tumors. Anti-androgens reduce androgen levels in targeted tissues [2]. PC is treated with a variety of approaches (single and combination therapy) including hormonal treatment, radiation implant, surgical removal of the prostate and/or chemical castration in order to eliminate testosterone production. In the initial stage of prostate cancer treatment, hormonal therapy is used primarily to target the androgen receptor (AR). Bicalutamide, Nilutamide, and Flutamide are recently available for the treatment of PC [3]. The antiandrogen inhibits the binding of androgen hormone to its receptor by either enhancing the apoptosis of cells or inhibiting the growth of prostate cancer cells [4, 5].

Flutamide is a nonsteroidal antiandrogen drug used to treat prostate cancer. Chemically, it is 4-nitro-3-trifluoromethyl-isobutylanilide (Figure 13.1). As the active metabolite of flutamide (metabolized by CYP1A2 in the liver), hydroxyflutamide can induce a mitogen activated protein kinase (MAPK) cascade in AR negative cells, thereby enhancing Cyclin D1 expression and cell proliferation. A neutralizing antibody against endothelial growth factor receptor (EGFR) had also blocked it [6–8].

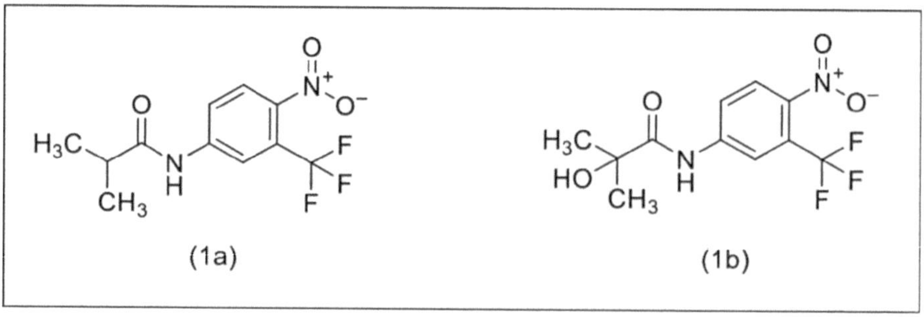

FIGURE 13.1 Structure of flutamide (1a); and hydroxyflutamide (1b).

As a high dose (750–1,500 mg/day), flutamide causes hepatotoxicity within three months from the treatment days, causing liver failure and photosensitivity reactions with erythema and vesiculobullous necrolytic lesions (rarely) [9–11]. Flutamide may cause side effects such as diarrhea, hematological alteration, muscle cramps, and gynecomastia. Flutamide also caused disorders like bleeding in the rectum, inflamed prostate, blood in the urine, liver malfunction, and methemoglobinemia [12].

Bioisosters are groups of atoms or molecules with the same physical and chemical properties, and they can be interchanged. Approaches of isosteric replacement of functional group are used to increase potency, decrease side effects, separate biological activity, and increase the duration of action by altering metabolism. They are classified as classical (Atoms and groups which have the same number of valence electrons) and non-classical Bioisosters (show a similarity in some important parameters to a certain extent) [17].

In the present study, the Bioisosteric approach has been used for the isopropyl group in the Flutamide molecule (Figure 13.2) in order to develop newer analogs with fewer side effects like hepatotoxicity, which is a major cause of drug withdrawal. Therefore, isopropyl bioisosteres of Flutamide are designed using the Bioisosteric approach to develop relatively safe compounds.

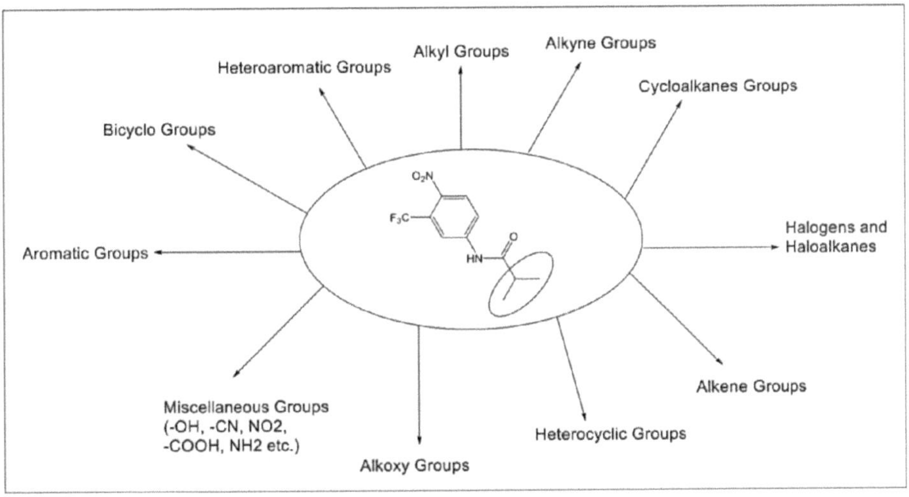

FIGURE 13.2 Bioisosters of isopropyl group in flutamide.

13.2 EXPERIMENTAL METHODS AND MATERIALS

13.2.1 DESIGN OF ISOPROPYL BIOISOSTERS OF FLUTAMIDE

A nonsteroidal antiandrogen, flutamide works by selectively blocking the AR. As shown in Figure 13.2, isopropyl bioisosteres of Flutamide were designed using MolOpt [13], and the molecular structure of newly designed analogs and their properties are listed in Table 13.1. ACD/ChemSketch 2021.1.1 [14] was used to draw the structure of ligands. The MolOpt tool is used for the *in-silico* design of newer flutamide analogs using bioisosteric approach by replacing the isopropyl group.

13.2.2 PHARMACOKINETIC (ADME) AND TOXICITY PROPERTIES PREDICTION

Various absorption, distribution, metabolism, excretion, and toxicity (ADMET) properties were calculated by using ADMETLab 2.0 online tool [15].

13.2.3 DRUG-LIKENESS (DL) AND DRUG-SCORE (DS)

Toxicity, drug likeness (DL) and drug score (DS), etc., were calculated by using Osiris property explorer (PEO) [16].

13.3 RESULTS AND DISCUSSION

13.3.1 BIOISOSTERS OF ISOPROPYL GROUP OF FLUTAMIDE

A total of 77 bioisosteres of the Isopropyl group were generated, such as n-Propyl, Trifluoromethyl, Cyclobutyl, Oxane, Oxolane, Morpholine, Diethylamino, Ethyl-propyl, Methyl Cyclopentyl, Methyl Cyclopentyl, etc., using the MolOpt tool. Some of these bioisosteres are shown in Table 13.1 based on QED, DILI, DL, and DS scores. Furthermore, these bioisosteres were investigated for ADMET prediction using the ADMETLab 2.0 online tool. The Lipinski rule of five, used in drug design and development, helps predict the oral bioavailability of potential lead molecules. The prediction of molecular properties, such as molecular weight (MW), number of hydrogen bond acceptors (nHA), number of hydrogen bond donors (nHD), and the logarithm of the n-octanol/water distribution coefficient (logP), for the newly designed analogs met the criteria of the Lipinski rule.

TABLE 13.1 Isopropyl Bioisosters of Flutamide and Molecular Properties of the Analogs

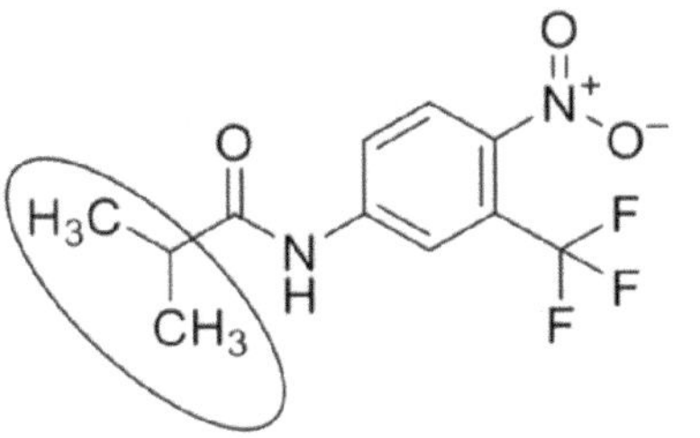

Flutamide

Sl. No.	Index	Bioisosters Type	LogS	LogD	LogP	MW	nHA	nHD	TPSA
1.	000	n-Propyl	–3.967	3.435	3.241	276.07	5	1	72.24
2.	002	Trifluoromethyl	–3.638	3.391	3.245	302.01	5	1	72.24
3.	007	Cyclobutyl	–4.855	3.529	3.313	288.07	5	1	72.24
4.	012	Cyclopropyl methyl	–4.461	3.452	3.355	288.07	5	1	72.24
5.	020	Ethyl-propyl	–5.122	4.072	3.941	304.1	5	1	72.24
6.	027	Iodo	–3.585	2.789	2.938	359.92	5	1	72.24
7.	028	Oxane	–4.017	3.007	2.72	318.08	6	1	81.47
8.	037	Oxolane	–3.687	2.839	2.463	304.07	6	1	81.47
9.	038	Morpholine	–3.216	2.613	2.37	319.08	7	1	84.71
10.	047	Diethylamino	–3.839	3.329	3.238	305.1	6	1	75.48
11.	049	Piperidine	–4.539	3.398	3.44	317.1	6	1	75.48
12.	050	Cyclopentyl methyl	–5.346	3.957	4.088	316.1	5	1	72.24
13.	051	Cyclobutyl methyl	–4.941	3.752	3.715	302.09	5	1	72.24
14.	054	Formaldehyde	–3.26	2.231	2.331	262.02	6	1	89.31
15.	055	Amino-methyl	–1.244	1.652	1.757	263.05	6	3	98.26
16.	076	Oxane	–3.976	3.166	3.194	318.08	6	1	81.47
17.	Flutamide	Isopropyl	–3.842	3.358	3.243	276.07	5	1	72.24

Abbreviations: MW: Molecular weight; nHA: number of hydrogen bond acceptor; nHD: number of hydrogen bond donor; nRot: number of rotatable bonds; logP: the logarithm of the n-octanol/water distribution coefficient; logS: the logarithm of aqueous solubility value; and TPSA: topological polar surface area.

13.3.2 PHARMACOKINETIC (ADME) AND TOXICITY PROPERTIES PREDICTION

Pharmacokinetic properties like absorption, distribution, metabolism, excretion (ADME) and toxicity properties were calculated by using ADMELab 2.0.

Medicinal Properties of the analogs are shown in Table 13.2. All compounds except 0.02, 027, and 054 have QED value (measure of drug-likeness) more than 0.67, indicating that these compounds are attractive. Synthetic accessibility of all compounds is excellent, that is ≤6. Lipinski, GSK, and golden triangle rules of all compounds have been accepted, which is the same as flutamide. Pfizer rule was followed by more than 50% of the compounds, whereas flutamide was rejected.

TABLE 13.2 Medicinal Properties of the Analogs

Sl. No.	Index	QED	Synth	Fsp3	MCE-18	Lipinski	Pfizer	GSK	Golden Triangle
1.	000	0.677	2.03	0.364	11	Accepted	Rejected	Accepted	Accepted
2.	002	0.518	2.283	0.222	14	Accepted	Rejected	Accepted	Accepted
3.	007	0.684	2.037	0.417	37.059	Accepted	Rejected	Accepted	Accepted
4.	012	0.682	2.123	0.417	35.294	Accepted	Rejected	Accepted	Accepted
5.	020	0.66	2.234	0.462	12	Accepted	Rejected	Accepted	Accepted
6.	027	0.288	2.608	0.125	11	Accepted	Accepted	Accepted	Accepted
7.	028	0.686	2.227	0.462	37.895	Accepted	Accepted	Accepted	Accepted
8.	037	0.687	2.72	0.417	52.059	Accepted	Accepted	Accepted	Accepted
9.	038	0.67	2.192	0.417	37.059	Accepted	Accepted	Accepted	Accepted
10.	047	0.683	2.219	0.417	12	Accepted	Accepted	Accepted	Accepted
11.	049	0.668	2.104	0.462	37.895	Accepted	Accepted	Accepted	Accepted
12.	050	0.67	2.145	0.5	37.143	Accepted	Rejected	Rejected	Accepted
13.	051	0.68	2.14	0.462	36.316	Accepted	Rejected	Accepted	Accepted
14.	054	0.389	2.366	0.111	11	Accepted	Accepted	Accepted	Accepted
15.	055	0.637	2.164	0.222	11	Accepted	Accepted	Accepted	Accepted
16.	076	0.686	2.706	0.462	52.895	Accepted	Accepted	Accepted	Accepted
17.	Flutamide	0.680	2.070	0.364	12	Accepted	Rejected	Accepted	Accepted

Abbreviations: QED: A measure of drug-likeness based on the concept of desirability; SAscore: synthetic accessibility score; Fsp3: the number of sp3 hybridized carbons/total carbon count; and MCE-18: medicinal chemistry evolution in 2018.

Absorption properties (Caco-2, MDCK, HIA) and distribution properties (plasma protein binding, volume of distribution, blood-brain barrier (BBB) and Fu) are shown in Table 13.3. All the analogs have proper caco-permeability and excellent MDCK. Human intestinal absorption of analog 028 and 076 has good oral absorption than flutamide, indicating that these compounds may be orally effective. Analogs 038 and 055 have excellent BBB results than flutamide, which means these analogs may not enter the brain. Analogs such as 038, 054, and 055 have proper PPB (less than 90%), whereas flutamide shows more than 90%. All analogs have an excellent volume of distribution.

TABLE 13.3 Absorption, Distribution, Metabolism, and Excretion Profile of the Analogs

Index	Caco-2	MDCK	HIA	BBB	PPB (%)	VDss	Fu (%)	CYP1A2	CYP2D6	CL	T1/2
000	–4.319	Ex.	0.004	0.60	95.49	0.72	5.78	–	–	7.58	0.33
002	–4.42	Ex.	0.005	0.51	96.56	0.90	2.89	+	–	44	0.13
007	–4.571	Ex.	0.005	0.58	96.17	0.82	3.85	+	–	3.22	0.16
012	–4.572	Ex.	0.005	0.72	95.54	0.69	5.05	–	+	6.43	0.16
020	–4.408	Ex.	0.004	0.53	96.63	0.79	3.07	+	+	3.96	0.18
027	–4.396	Ex.	0.005	0.54	93.26	0.74	5.58	+	+	2.78	0.25
028	–4.599	Ex.	0.003	0.34	92.28	0.78	6.59	–	–	7.56	0.24
037	–4.608	Ex.	0.004	0.31	91.40	0.77	6.42	–	+	5.72	0.29
038	–4.579	Ex.	0.004	0.25	84.68	0.94	9.46	+	+	6.92	0.59
047	–4.42	Ex.	0.004	0.82	93.57	0.92	5.24	–	+	8.98	0.33
049	–4.603	Ex.	0.005	0.78	95.09	0.94	4.89	–	+	5.32	0.2
050	–4.609	Ex.	0.004	0.55	97.26	0.76	2.37	+	+	5.86	0.10
051	–4.58	Ex.	0.004	0.64	96.44	0.73	3.47	+	+	6.24	0.13
054	–4.437	Ex.	0.008	0.56	82.75	1.02	14.06	+	+	2.04	0.28
055	–4.936	Ex.	0.007	0.23	53.10	1.24	34.10	–	–	8.01	0.76
076	–4.563	Ex.	0.003	0.44	95.37	0.76	4.33	+	+	4.45	0.21
Flutamide	–4.346	Ex.	0.004	0.58	95.62	0.83	4.59	+	+	4.68	0.24

Abbreviations: Caco-2: The human colon adenocarcinoma cell lines; MDCK: Madin–Darby canine kidney cells and >2 × 10^{-6} cm/s indicates excellent; HIA: human intestinal absorption; PPB: plasma protein binding; BBB: blood-brain barrier; VD: volume distribution; Fu: the fraction unbound in plasma; (–) indicates inhibitor; and (+) indicates substrate (human cytochrome P450 isozymes)—1A2 and 2D6); CL: the clearance of a drug; $T_{1/2}$: the half-life of a drug; and Ex.: excellent.

Metabolism (CYP1A2-inh and sub, CYP2D6-inh) and excretion (CL and $T_{1/2}$) properties are shown in Table 13.3. More than 50% Compounds have clearance value (CL) more than 5, suggesting excellent clearance.

Toxicity properties such as H-HT, DILI, AMES Toxicity, NR-AR are shown in Table 13.4. The results indicate that all analogs have lesser human hepatotoxicity (H-HT) than the standard drug. Analogues have drug-induced liver injury (DILI) less than the standard drug except 002, 038, 047, and 049. Compounds 002 and 047 have no AMES toxicity, which is less than 0.3 (No risk for mutagenicity), and compounds 007 and 020 have lower toxicity than flutamide. All analogs showed interactions with the androgen receptor (NR-AR) more than 0.7 (high risk), which is desirable for anti-androgen activity. Interactions of the compound with AR are crucial in prostate cancer. The results indicate that compound 020 and 055 may be the good lead on the basis of QED value (> 0.67) and low toxicity (H-HT and DILI) as compared to Flutamide. This study could lead to the development of a potent AR antagonist for the treatment of prostate cancer.

TABLE 13.4 Toxicity Profile, DL, and DS Prediction of the Analogs

Index	H-HT	DILI	Ames	Carc.	Res.	NR-AR	NR-AR-LBD	Toxicity Risk				DL	DS
								M	T	I	R		
000	0.411	0.691	0.573	0.766	0.858	0.832	0.071	G	G	G	G	−15.1	0.42
002	0.577	0.866	0.131	0.741	0.969	0.84	0.27	G	G	G	G	−33.2	0.41
007	0.401	0.667	0.666	0.777	0.931	0.844	0.056	G	G	G	G	−12.3	0.41
012	0.556	0.463	0.817	0.827	0.893	0.842	0.082	G	G	G	G	−11.8	0.41
020	0.417	0.492	0.318	0.698	0.918	0.871	0.011	G	G	G	G	−10.6	0.38
027	0.2	0.583	0.779	0.848	0.957	0.798	0.301	G	G	G	G	−13.3	0.4
028	0.34	0.653	0.775	0.845	0.506	0.764	0.036	G	G	G	G	−15.1	0.42
037	0.328	0.574	0.666	0.806	0.61	0.789	0.201	G	G	G	G	−13.4	0.43
038	0.323	0.925	0.848	0.892	0.108	0.752	0.031	G	G	G	G	−11.4	0.44
047	0.306	0.911	0.234	0.686	0.879	0.82	0.036	G	G	G	G	−10.5	0.42
049	0.324	0.859	0.641	0.822	0.833	0.839	0.023	G	G	G	G	−12.4	0.4
050	0.511	0.496	0.815	0.722	0.928	0.841	0.033	G	G	G	G	−14.6	0.37
051	0.519	0.454	0.813	0.782	0.917	0.838	0.053	G	G	G	G	−13.2	0.39
054	0.185	0.58	0.866	0.856	0.954	0.78	0.597	O	G	G	G	−15.2	0.37
055	0.344	0.425	0.655	0.678	0.828	0.798	0.467	G	G	G	G	−13.2	0.46
076	0.439	0.645	0.688	0.653	0.83	0.743	0.179	G	G	G	G	−17.0	0.41
Flutamide	0.578	0.858	0.498	0.786	0.95	0.862	0.018	R	O	G	O	−12.9	0.16

Abbreviations: H-HT: The human hepatotoxicity; DILI: drug-induced liver injury; Ames: test for mutagenicity; Carc.: carcinogenic; Res.: respiration; ROA: rat oral acute toxicity; NR-AR: androgen receptor-a nuclear hormone receptor; NR-AR-LBD: molecule bind with LBD of androgen receptor bind; M: mutagenic; T: tumorigenic; I: irritant; R: reproductive; G: no toxicity risk; O: toxicity risk; R: high toxicity risk; DL: drug likeness; and DS: drug score.

13.3.3 DRUG-LIKENESS (DL) AND DRUG-SCORE (DS) PREDICTION

Drug-likeness (DL) and drug-score (DS) are properties of molecules that indicate their attractiveness and are calculated by using Osiris Property explorer as shown in Table 13.4. Some compounds have good DL value (compound 020 and 055) and DS value (compound 055) than Flutamide (standard drug).

13.4 CONCLUSION

Flutamide analogs (bioisosters of isopropyl group) were designed by bioisosteric approach and ADMET properties calculated by different computational tools. The objective of the *in-silico* design was to develop less toxic analogs and more potent than the standard drug (Flutamide). Analogs 007 and 020 are prominent analogs on the basis of less toxicity (DILI <0.7), exhibit drug-likeness (QED ≥ 0.67), and interact well with the androgen receptor (NR-AR ≥ 0.7). On the other hand, analogs such as 0,37,055 and 076 also showed less toxicity than Flutamide. Therefore, these compounds may have great potential as a drug candidate for the treatment of prostate cancer. In the future, analogs of flutamide (compound 007 and 020) could be used as a lead compound.

KEYWORDS

- **ADMET**
- **anti-androgen**
- **bioisosteric approach**
- **drug score**
- **flutamide**
- **isopropyl group**

REFERENCES

1. International Agency for Research on Cancer, (2020). *India Fact Sheet 2020: Globocan.* https://gco.iarc.fr/today/data/factsheets/populations/356-india-fact-sheets.pdf (accessed on 21 December 2023).
2. Stabile, R. G., & Dicks, A. P., (2003). Microscale synthesis and spectroscopic analysis of flutamide, an antiandrogen prostate cancer drug. *J Chem Educ.*, *80*(12), 1439–1443.

3. He, H., Morely, J. E., Silva-Lopez, E., Bottenus, B., Montajano, M., Fugate, G. A., Twamley, B., & Benny, P. D., (2009). Synthesis, and characterization of nonsteroidal-linked $M(CO)^{3+}$ ($M = {}^{99m}Tc$, Re) compounds based on the androgen receptor targeting molecule flutamide. *Bioconjug Chem.*, *20*(1), 78–86.
4. Yamini, Y., Kalantarian, P., Hojjati, M., Esrafily, A., Moradi, M., Vatanara, A., & Harrian, I., (2010). Solubilities of flutamide, dutasteride, and finasteride as antiandrogenic agents, in supercritical carbon dioxide: Measurement and correlation. *J. Chem. Eng. Data.*, *55*, 1056–1059.
5. Pucci, E., Genazzani, A. D., Monzani, F., Lippi, F., Angelini, F., Gargani, M., Barletta, D., Luisi, M., & Genazzani, A. R., (1995). Prolonged treatment of hirsutism with flutamide alone in patients affected by polycystic ovary syndrome. *Gynecol. Endocrinol.*, *9*(3), 221–228.
6. Mehrabi, A., Rahimnejad, M., Mohammadi, M., & Pourali, M., (2019). Electrochemical detection of flutamide with gold electrode as an anticancer drug. *Biocatal. Agric. Biotechnol.*, *22*, 101375.
7. Kostrubsky, S. E., Strom, S. C., Ellis, E., Nelson, S. D., & Mutlib, A. E., (2007). Transport, metabolism, and hepatotoxicity of flutamide, drug-drug interaction with acetaminophen involving phase I and phase II metabolites. *Chem Res Toxicol.*, *20*(10), 1503–1512.
8. Limaye, A. M., Asangani, I., Bora, N., & Kondaiah, P., (2007). Novel flutamide regulated genes in the rat ventral prostate: Differential modulation of their expression by castration and flutamide treatments. *Asian J. Androl.*, *9*(6), 801–808.
9. Castelo-Branco, C., & Pino, M. D., (2009). Hepatotoxicity during low-dose flutamide treatment for hirsutism. *Gynecol Endocrinol.*, *25*(7), 419–422.
10. Hejmej, A., & Bilinska, B., (2018). The effects of flutamide on cell-cell junctions in the testis, epididymis, and prostate. *Reprod. Toxicol.*, *81*, 1–16.
11. Kojima, M., Kamoi, K., Ukimura, O., Fujito, A., Nakao, M., Tanaka, S., Miyashita, H., et al., (2002). Clinical utility of ursodeoxycholic acid in preventing flutamide-induced hepatopathy in patients with prostate cancer: A preliminary study. *Int J Urol.*, *9*, 42–46.
12. Kubendhiran, S., Sakthivel, R., Chen, S. M., Mutharani, B., & Chen, T. W., (2018). Innovative strategy based on a novel carbon-black-β-cyclodextrin nanocomposite for the simultaneous determination of the anticancer drug flutamide and the environmental pollutant 4-nitrophenol. *Anal Chem.*, *90*(10), 6283–6291.
13. Shan, J., & Ji, C., (2020). MolOpt: A web server for drug design using bioisosteric transformation. *Curr. Comput. Aided Drug Des.*, *16*(4), 460–466.
14. *ACD/Chem Sketch*, (2021.1.1). ACD/labs software and services. https://www.acdlabs.com (accessed on 21 December 2023).
15. Xiong, G., Wu, Z., Yi, J., Fu, L., Yang, Z., Hsieh, C., Yin, M., et al., (2021). ADMETlab 2.0: An integrated online platform for accurate and comprehensive predictions of ADMET properties. *Nucleic Acids Res.*, *49*, W5–W14.
16. Thomas, S., (2001). *OSIRIS Property Explorer.* Organic chemistry portal. https://www.organic-chemistry.org/prog/peo/ (accessed on 21 December 2023).
17. Kumari, S., Carmona, A. V., Tiwari, A. K., & Trippier, P. C., (2020). Amide bond bioisosteres: Strategies, synthesis, and successes. *J. Med. Chem.*, *63*(21), 12290–12358.

CHAPTER 14

A REVIEW ON MECHANICAL PROPERTIES OF COIR REINFORCED POLYMER COMPOSITES

VIVEK BORKAR[1] and DILIP MISHRA[2]

[1]*Mechanical Engineering Department, Rungta College of Engineering and Technology, Bhilai, Chhattisgarh, India*

[2]*Faculty of Science and Technology, ICFAI University, Raipur, Chhattisgarh, India*

ABSTRACT

Coir reinforced polymer composites are an exciting area of material science that combines the natural strength of coir fibers with the versatility of polymer matrices to create materials with enhanced mechanical properties. These composites are noted for their improved strength compared to those made from polymers alone. By integrating coir fibers into polymer substrates, researchers have been able to significantly enhance the tensile strength of the resulting composite material. The mechanical properties of coir reinforced polymer composites are a subject of considerable interest due to their potential applications in various industries, including automotive, construction, and aerospace. The tensile strength, a critical mechanical property, is markedly improved in these composites due to the strong interfacial bonding between the coir fibers and the polymer matrix. This improvement in tensile strength is a key factor in the suitability of coir reinforced polymer composites for load-bearing applications. In addition to tensile strength, these composites also exhibit enhanced flexural strength and stiffness, making them more resistant to bending and deformation

Computational Optimization, Modeling, and Simulation for Engineering Applications.
Anupam Shukla, Sourabh Rungta, Mohan Awasthy, and Rakesh Himte (Eds.)

under stress. The impact resistance of coir reinforced polymer composites is another mechanical property that benefits from the incorporation of coir fibers, offering better durability and longevity. The mechanical properties of coir reinforced polymer composites can be further tailored by modifying the fiber content, orientation, and the type of polymer matrix used. This customization allows for the optimization of composites for specific applications, maximizing their performance and efficiency. The improved properties of these composites, such as their enhanced strength-to-weight ratio, make them an attractive alternative to traditional materials. Sustainability is another advantage of coir reinforced polymer composites. Coir, derived from coconut husks, is a renewable resource that contributes to the environmental friendliness of these composites. By utilizing natural fibers like coir, the manufacturing of polymer composites becomes more sustainable, reducing the dependency on synthetic, non-renewable fibers. The biodegradability of coir also adds to the eco-friendliness of these composites, making them suitable for applications where environmental impact is a concern. In this chapter, coir reinforced polymer composite's improved mechanical properties such as tensile strength, impact resistance, and stiffness have been reported and comparison of Coir reinforcement in different matrix with its properties have been presented.

14.1 INTRODUCTION

In the past decade, there have been considerable efforts in reducing global warming and the use of natural products will help in countering this situation by increasing their demand which will eventually increase their production, and hence the green cover will increase. In this regard, coir fiber can come up as a good option for reinforcement as its usage has been reported to increase the tensile strength, flexural strength when reinforced in Polypropylene powder [1].

Also, in another such research, it was found that by using coir fiber as reinforcement, the tensile strength of the unreinforced polyvinyl alcohol (PVA) matrix increased by 96.95% [2]. Additionally, coir fiber is versatile and low in cost in tropical countries like India [3]. Furthermore, using coir of different grain sizes shows improvement in various mechanical properties of the polymer composites [4]. It has also been found that using hybrid reinforcement with coir, such as titanium carbide (TiC), improves the mechanical properties of the polymer composites [5]. Moreover, it is being used with other natural fibers like basalt innegra fiber, along with titanium carbide (TiC), to improve thermal stability and mechanical properties [6]. Many methods have been found and tested for using coir as reinforcement in polymer composites, which result in improved mechanical and thermal properties compared to the unreinforced ones [1–4, 6–12].

Coir is a natural fiber that is easily available in India, and its production is not very harmful to nature. Additionally, the production process requires many semi-skilled and unskilled workers, which can help solve the low employment rate in the country. Due to these benefits, coir could be used instead of glass fiber, which is more expensive, in making thermoplastics [1]. The coir fiber reinforced polymer can be used to replace the existing interior of automobiles, which is mainly made of plastic and other pollution-causing materials, because coir has better mechanical and thermal properties [1]. It has also been found that coir has better properties than bagasse, which is another available natural fiber [2]. It has been found that the geometry of natural fiber composites plays an important role in determining the mechanical properties of the polymer composite [11]. Therefore, further testing is needed to optimize the properties of composites by testing different fiber geometries according to the requirements.

Using red mud as a filler material with other abaca fibers has proven to improve the mechanical properties of the natural fiber composite materials, so such experimentation needs to be done with coir to check how it changes its properties [13]. Several review articles are available that deal with the mechanical, chemical, and thermal properties of the coir composite [14], but do not discuss the individual tensile strength or flexural strength, which actually indicate the changes produced in the composites. In this review chapter, individual discussions about tensile strength, flexural strength, and impact strength have been provided to gain a better understanding of the mechanism of strengthening of coir fiber polymer composites.

There are a limited number of research available which deal with coir fiber as the reinforcement in polymer composites [15]. Thus, there is a need for such a study which would provide the benefits of using coir fiber as the reinforcement material. And this would be fulfilled by the present study. This research paper deals with the mechanical properties of coir fiber reinforced polymer composites wherein tensile strength, flexural strength, and impact strength are discussed separately along with the factors affecting them. Moreover, research gaps are also unveiled which would be beneficial for researchers and scientists working in this domain.

14.2 MECHANICAL PROPERTIES OF COIR REINFORCED POLYMER COMPOSITE

14.2.1 TENSILE STRENGTH

Tensile property is an important mechanical property as it gives important information regarding the usage of the material. Many efforts have been carried out to study the tensile property of coir reinforced composite polymer. In one such study [1], a coir fiber reinforced polymer composite material is prepared in a polypropylene matrix. It is grafted with 3 wt.% maleic anhydrides to act as a coupling agent. In this study, the coir fiber reinforcement weight percentage is gradually increased, and it is seen that by increasing its weight percentage to about 70%, there is an increase in tensile strength by 26%. Another research paper has evaluated cellulose nanocrystals of coir. These are obtained using mercerization, acid hydrolysis, and bleaching in a PVA-based matrix. It showed that in a 0.5% weight coir composite, there was a 96.9% increase in tensile strength compared to the unreinforced polymer matrix [2].

But not all such research shows an increase in tensile strength with an increase in weight percentage of coir fiber in the polymer matrix. In one such paper in which polylactic acid is used as a composite and maleic anhydride grafted PLA (PLA-g-MA) as a compatibilizer. In this study, the coir fiber weight percentage is increased from 10 to 30. Its tensile strength is seen to decrease from 57.4 MPa to 52.8 MPa. The reason for this, as found by acid titration, was an increase in acidity of the polymer due to the compatibilizer and a reduction in the moisture content of the fiber. Coir fiber can also be used with other materials as hybrid reinforcement to check its effect on the tensile property of the reinforced matrix. In a research, TiC reinforced coir fiber was used in two types of epoxy resins: bio-epoxy Sr 33 (GreenPoxy (GP33) with Surf Clear ultra-slow hardener SD 4990) and synthetic epoxy Epotec YD-535 L. It was seen that reinforcement of GreenPoxy GP33 Reinforced with both coir fiber and TiC Nanoparticles (5% coir: 5% TiC) showed an improvement in tensile strength of 10.26% and 46.27% [5] compared to both pure unreinforced matrix. Tensile strength of the reinforced coir fiber depends on many factors of the reinforced matrix composite preparation and combination. One such factor is fiber length. In a research paper, epoxy matrix was reinforced with coir fiber and the length of coir fiber was increased. It was seen that by increasing the length of coir from 5 mm to 30 mm in 70% epoxy by weight, tensile strength increased from 3.2 MPa to 13.05 MPa [9]. Treatment of coir fiber is another variable that can be varied, and its effect can be observed for changes in tensile strength properties. In one such research, three different treatments were used for polylactic acid-based coir reinforced polymer composite, which are 2% Silane, 6% Ca(OH)2, and 6% NaOH. It showed that Untreated Coir Fiber had the highest tensile strength of 104.8 MPa, followed by 74.5 MPa (2% Silane), 69.9 MPa (6% Ca(OH)2), and 55.5 MPa (6% NaOH) for treated fibers [10]. In another research, the geometry of the coir fibers in the epoxy matrix and silicon carbide was used as the filler material. It was changed in 3 ways: Unidirectional, Woven ±45° orientation, and chopped. Changes observed in tensile strength of the composite gave results that Unidirectional had the highest 21 MPa of tensile strength than the other 2 configurations of coir fiber [11].

14.2.2 FLEXURAL STRENGTH

Flexural strength is an important property as it gives the bending strength of the polymer composite material which helps in deciding its application. Many studies have been carried out to know changes in the Flexural property of polymer under different combinations of fiber composition, fiber treatment, and fiber treatment. In a study, the weight percentage of the coir fiber is increased in polypropylene matrix having a coupling agent as 3 wt.% maleic anhydrides in it, the Flexural strength is observed to increase from 35% by increasing the weight percentage from 40% to 70% [1]. But it cannot be assumed here that it is a general case that by increasing the weight percentage of the coir fiber in the polymer composite increases the Flexural strength of the polymer as it can be seen in the next chapter in which malic anhydride is used as a compatibilizer and polylactic acid as the matrix and coir as reinforcement shows a decrease in Flexural strength of the composite from 99.8 MPa to 95.5 MPa with an increase in weight percentage of the coir fiber due to an increase in acidity in the polymer due to the compatibilizer [3]. Till now it is seen that the weight percentage of coir was changed in the composite polymer to observe a change in Flexural strength of the composite but the grain size of the fiber can also be varied and its effect on Flexural strength was observed in this chapter in which epoxy composite was used, weight percentage was kept at 10% and an alkali solution of sodium hydroxide (NaOH) solution with 2.0% w/v for 3 h was used [3]. The grain size when increased from 250 μm to 950 μm showed a considerable increase in Flexural strength of about 94%. In another paper, two types of matrices were used bio-epoxy Sr 33 (GreenPoxy (GP33) with SurfClear ultra-slow hardener SD 4990) and synthetic epoxy Epotec YD-535 L and hybrid reinforcement of coir with TiC was done [6]. The addition of TiC nanoparticles within the bio-matrix improved the Flexural strength from

115.05 MPa to 124 MPa [6]. Also, it was seen that Flexural strength in the case of coir epoxy composite was also enhanced. Coir fiber length is also a variable in making a reinforced polymer composite, it can be varied to check its effect on the Flexural properties of the composite. In a research paper, the length of the coir fiber was increased from 5 mm to 30 mm in (coir: epoxy: 30:70% by weight) and an improvement in Flexural strength was observed from 25.41 MPa to 35.42 MPa [6]. By changing the configuration of coir fiber polymer composite, its Flexural properties can be varied. In one such research having coir fiber reinforcement in the SiC particle-filled epoxy composite, 3 types of configurations are used: unidirectional, woven ±45° orientation, and chopped, and it is observed that Flexural strength is maximum for unidirectional, i.e., 21 MPa, values for Woven ±45° orientation is 40 MPa and chopped is 29 MPa [11].

14.2.3 IMPACT STRENGTH

Impact strength is an important property of the material as it gives us the application where material can be used, so we test coir reinforced composites for their impact strength as well and try to discover how their impact strength is affected by different changes in weight, composition, treatment of the composite. In such a research paper, it is seen that by decreasing the grain size of the coir from 950 μm to 250 μm in the epoxy matrix treated with alkaline solution of NaOH increases the impact strength gradually from 5 kJ/m^2 to 9 kJ/m^2[4] also all the reinforced composites show improvement in the impact strength compared to the unreinforced epoxy matrix. The reason for reinforced composites showing better impact strength lies in the fact that there is good adhesion between the fiber and the matrix, which diminishes the crack propagation. However, a decrease in impact strength with an increase in grain size is due to the inability of the reinforcement to stop the propagation of the crack with an increase in grain size of coir fiber [4]. To further investigate the scope of improvement of impact strength of coir reinforced composites, a research paper used two types of matrix, i.e., bio-epoxy Sr 33 (GreenPoxy (GP33) with Surf Clear ultra-slow hardener SD 4990) and synthetic epoxy Epotec YD-535 L. Also, hybrid reinforcement was used with coir of TiC, and results showed that in both types of matrix, hybrid reinforcement showed improvement in impact strength [6]. In a research paper using epoxy matrix in 70% by weight and coir fiber was reinforced into it, its length was changed from 5 mm to 30 mm. and impact strength was improved from 16 kJ/m^2 to 17.5 kJ/m^2[9]. Table 14.1 tabulates the mechanical properties of various coir fiber reinforced polymer composites.

TABLE 14.1 Mechanical Properties of Coir Reinforced Polymer Composites

Reinforcements	Type of Matrix	Wt.% of Reinforcement	Treatment of Coir	Tensile Strength (MPa)	Flexural Strength (MPa)	Impact Strength (kJ/m2)	References
Coir	PP	40	–	13.2	24.3	–	1
Coir	PP	50	–	15.5	26.8	–	
Coir	PP	60	–	17.8	30.6	–	
Coir	PP	70	–	16.0	27.3	–	
Coir	PVA	0.5	2 wt.% NaOH at 90?C for 16 h.	38.2	–	–	2
Coir	PVA	2	2 wt.% NaOH at 90?C for 16 h.	32.8	–	–	
Coir	PLA	10	–	57.4	99.8	–	3
Coir	PLA	30	–	52.8	95.5	–	
Coir	PLA	10	–	56.9	104.8	–	
Coir	PLA	10	–	57.9	107.1	–	
Coir	PLA	30	–	45.5	73.5	–	
Coir	PLA	30	–	45.2	83.9	–	
Coir	Epoxy resin	–	Sodium hydroxide (NaOH) solution with 2.0% w/v for 3 h.	–	103.57	5.45	4
Coir: TiC nanoparticles	CABE33	Coir: 5% TiC: 5%	–	31	125	4.5	5
Coir	Epoxy resin	Coir: 30% Fiber length: 30 mm	–	13.05	35.42	17.5	6
Coir	Polymer composite	–	Sodium hydroxide (NaOH) 6% w/v	55.5	–	–	7

TABLE 14.1 *(Continued)*

Reinforcements	Type of Matrix	Wt.% of Reinforcement	Treatment of Coir	Tensile Strength (MPa)	Flexural Strength (MPa)	Impact Strength (kJ/m2)	References
Coir	Polymer composite	–	Calcium hydroxide $Ca(OH)_2$ 6% w/v.	69.9 MPa	–	–	
Coir	Polymer composite	–	Silane (triethoxy(ethyl) silane) 2%	74.5 MPa	–	–	
Coir:SiC	Epoxy	10 wt.% of coir fiber with 7–8 wt.% of silicon carbide.	–	14.02	29	–	8
Coir	Epoxy	25 wt.% of coir 75 wt.% of epoxy.	5% aqueous NaOH solution for 3 hours at room temperature.	33.43	65.07	65.57 (j/m)	9
Coir/wool/gypsum	–	20/10/70	–	–	3.78	–	10

Note: PP: Polypropylene; CNC: cellulose nanocrystal; PVA: polyvinyl alcohol; PLA: poly(lactic acid); GreenPoxy GP33 reinforced with both coir fiber and TiC nanoparticles-CABE33.

14.3 CONCLUSION

When natural fibers are used as reinforcement in polymer composites to achieve improvements in mechanical and thermal properties, another problem is solved: the waste of these natural fibers. Farmers can now earn value from these fibers, which were previously wasted and did not provide any benefits for them. This can help increase the income for farmers. Additionally, many waste materials like fly ash can also be used for reinforcement, which reduces the problem of safe disposal in the environment. This review analyzes such natural fiber composites to determine changes in their mechanical properties with different configurations. The goal is to find the configuration that results in the most improvements in mechanical properties. Furthermore, this research can help identify missing links in the configuration of natural fiber polymer composites. These missing links can be tested to determine if they have not been tested yet or if there is limited available work on them. However, they have the potential to provide good mechanical properties if tested.

It has been observed from literature that increasing the content of coir fiber in the composite can improve the tensile strength and flexural strength. But, on increasing the coir fiber concentration beyond a critical limit, the tensile strength decreases. The reason for this decrement is that the polymer is not able to properly hold the fibers. Thus, upcoming researchers can take this as a challenge to further improve the strength of the composite. Coir fiber can also be used as cellulose nanocrystals, and these crystals show higher bonding with the poly-vinyl alcohol using hydroxyl bond than bagasse, which leads to an improvement in mechanical properties. Also, the total crystallinity index of these cellulose nanocrystals of coir is higher than that of bagasse. Again, on increasing cellulose nano-crystal concentration beyond a critical limit, the tensile strength reduces due to the formation of agglomerates. Therefore, future researchers can solve this problem to give better improvement in tensile strength. The tensile strength of the polymer is improved due to reinforcement only when the interfacial interaction between the polymer and fiber is more. When it is less, this may lead to a reduction in the tensile strength of the polymer. Flexural strength and impact strength are two important mechanical properties that are analyzed in this chapter. For an overall improvement in the mechanical property of a composite, both properties must improve. But it is seen that by increasing the grain size, the flexural strength increases but results in reducing impact strength. Hence, there is a scope for further work in which optimization can be done to balance the improvement in both properties.

To improve the tensile strength and impact strength of the composite, hybrid reinforcement can be used by using TiC as reinforcement with coir, which would lead to an improvement in tensile and impact strength. Moreover, there is a lot of scope for testing other fibers/particulates as reinforcement with coir to improve the mechanical properties using filler. An increase in fiber length in the composite reported an improvement in tensile strength, flexural strength, and impact strength. Chemical treatment of coir fiber is done to remove undesirable substances and to improve the adhesion between the matrix and fiber, but there was a decrement observed in the tensile strength of coir fiber after treatment with Silane, $Ca(OH)_2$ and NaOH. Reason for this is concluded to be thickening of cell wall, so there is a need to find a suitable treatment process that can improve mechanical properties of coir fiber. Moreover, coir fiber geometry also plays a significant role in determining the improvement of tensile strength of reinforced composites.

KEYWORDS

- **coir**
- **flexural strength**
- **impact strength**
- **mechanical properties**
- **polymer composites**
- **tensile strength**

REFERENCES

1. Ayrilmis, N., Jarusombuti, S., Fueangvivat, V., Bauchongkol, P., & White, R. H., (2011). Coir fiber reinforced polypropylene composite panel for automotive interior applications. *Fibers Polym., 12*(7), 919–926. doi: 10.1007/s12221-011-0919-1.
2. Pavalaydon, K., Ramasawmy, H., & Surroop, D., (2021). Comparative evaluation of cellulose nanocrystals from bagasse and coir agro-wastes for reinforcing PVA-based composites. *Environ. Dev. Sustain.*doi: 10.1007/s10668-021-01852-9.
3. Rigolin, T. R., Takahashi, M. C., Kondo, D. L., & Bettini, S. H. P., (2019). Compatibilizer acidity in coir-reinforced PLA composites: Matrix degradation and composite properties. *J. Polym. Environ., 27*(5), 1096–1104. doi: 10.1007/s10924-019-01411-4.

4. Sadeq, N. S., Mohammadsalih, Z. G., & Mohammed, R. H., (2020). Effect of grain size on the structure and properties of coir epoxy composites. *SN Appl. Sci., 2*(7). doi: 10.1007/s42452-020-2991-x.
5. Arshad, M. N., et al., (2021). Effect of coir fiber and TiC nanoparticles on basalt fiber reinforced epoxy hybrid composites: Physico–mechanical characteristics. *Cellulose, 28*(6), 3451–3471. doi: 10.1007/s10570-021-03752-7.
6. Mohit, H., et al., (2021). Effect of TiC nanoparticles reinforcement in coir fiber-based bio/synthetic epoxy hybrid composites: Mechanical and thermal characteristics. *J. Polym. Environ., 29*(8), 2609–2627. doi: 10.1007/s10924-021-02069-7.
7. Gupta, N., & Ramkumar, P. L., (2021). Experimental investigation of linear low density polyethylene composites based on coir for rotational molding process. *Polym. Polym. Compos., 29*(8), 1114–1125. doi: 10.1177/0967391120953246.
8. Hasan, K. M. F., et al., (2021). Novel insulation panels development from multilayered coir short and long fiber reinforced phenol formaldehyde polymeric biocomposites. *J. Polym. Res., 28*(12). doi: 10.1007/s10965-021-02818-1.
9. Sanjay, K., Amar, P., & Biswas, S., (2011). Effect of fiber length mechanical behavior of coir fiber reinforced epoxy composites. *Fibers Polym., 12*(1), 73–78. [Online]. Available: 10.1007/s12221-011-0073-9.
10. Siakeng, R., Jawaid, M., Ariffin, H., & Salit, M. S., (2018). Effects of surface treatments on tensile, thermal, and fiber-matrix bond strength of coir and pineapple leaf fibers with poly lactic acid. *J. Bionic Eng., 15*(6), 1035–1046. doi: 10.1007/s42235-018-0091-z.
11. Mishra, S., Nayak, C., Sharma, M. K., & Dwivedi, U. K., (2021). Influence of coir fiber geometry on mechanical properties of SiC filled epoxy composites. *Silicon, 13*(2), 301–307. doi: 10.1007/s12633-020-00425-1.
12. Mohit, H., et al., (2021). Nanoparticles addition in coir-basalt-innegra fibers reinforced bio-synthetic epoxy composites. *J. Polym. Environ., 29*(11), 3561–3573. doi: 10.1007/s10924-021-02133-2.
13. Sinha, K., Bhattacharya, S., & Narang, H. K., (2019). Experimental determination and modelling of the mechanical properties of hybrid abaca-reinforced polymer composite using RSM. *Polym. Polym. Compos., 27*(9), 597–608. doi: 10.1177/0967391119855843.
14. Adeniyi, G., Onifade, D. V., Ighalo, J. O., & Adeoye, A. S., (2019). A review of coir fiber reinforced polymer composites. *Composites Part B: Engineering, 176*. Elsevier Ltd. doi: 10.1016/j.compositesb.2019.107305.
15. Sinha, K., Narang, H. K., & Bhattacharya, S., (2017). Mechanical properties of natural fiber polymer composites. *J. Polym. Eng., 37*(9), 879–895. doi: 10.1515/polyeng-2016-0362.

CHAPTER 15

IN-SILICO DESIGN AND ADMET STUDY OF NOVEL ENTRECTINIB ANALOGS AS POTENT TRK INHIBITORS

B. G. V. S. JAGAN and SANMATI K. JAIN

Department of Pharmacy, Guru Ghasidas Vishwavidyalaya (A Central University), Bilaspur, Chhattisgarh, India

ABSTRACT

Entrectinib is a first-generation and potent tropomyosin receptor kinase (Trk/TRK) inhibitor showing activity on all 3 Trk-A, Trk-B, and Trk-C subtypes and is used as a promising therapy for advanced or recurrent TRK fusion-positive cancers. It is approved as a third "tissue agnostic" cancer drug against solid tumors and non-small cell lung cancer (NSCLC). The drug possesses a high toxicity level, and the present study is focused on *in-silico* design of less toxic and more potent analogs with fewer side effects than the standard drug (Entrectinib) by bioisosteric approach using MolOpt software and further modification of the bioisosteres based on ADMET results. ADMETLab 2.0 software is used to predict Pharmacokinetic (absorption, distribution, metabolism, and excretion) and toxicity properties for the newly designed analogs. The results indicate that the analogs 11, 15, and 19 are potent based on their least toxicity levels (DILI score < 0.3) and exhibiting potential drug-likeness (QED > 0.67). Therefore, these newly designed analogs may have the potential to be used as lead compounds if further found to deliver promising and reliable experimental results as potent TRK inhibitors.

Computational Optimization, Modeling, and Simulation for Engineering Applications.
Anupam Shukla, Sourabh Rungta, Mohan Awasthy, and Rakesh Himte (Eds.)

15.1 INTRODUCTION

Cancer is among the leading causes of mortality and is impacting barrier to life expectancy with an estimated 1.898 million new cancer cases and 0.6 million cancer-related deaths occurred in 2021 as per American Cancer Society statistics across the globe, making it a key area to work upon [1]. Kinases are a part of a large family of enzymes that catalyze the phosphorylation process, and if the substrate is a protein, then they are termed as protein kinases (PKs). Based on the specificity of amino acid, PKs may be serine-threonine specific kinase or protein-tyrosine kinase [2]. Most of the PKs share highly conserved catalytic domains and possess similarity in their sequence and structure. This leads to a lack of selectivity and off-target toxicity of drug moieties [3].

Tropomyosin receptor kinases (Trks/TRKs) belong to the class of Tyrosine-specific PKs, i.e., receptor tyrosine kinases (RTKs). The function of RTKs is to catalyze the process of transferring a phosphoryl group from a nucleoside triphosphate (a donor) to the hydroxyl group of tyrosine residues on protein substrates and thereby triggering the activation of downstream signaling cascades. Any kind of mutations, translocations, or amplifications can cause abnormal activation of tyrosine kinases and further implicate in tumorigenesis, invasion, progression, and metastasis of malignancies [4]. The design of a tyrosine kinase inhibitor (TKI) is such that it will inhibit the corresponding kinase in playing its role of catalyzing the phosphorylation process. By now, at least 58 RTKs and 32 non-receptor tyrosine kinases (NRTKs) have been reported. TRKs control synaptic strength, plasticity, neuronal survival, proliferation, migration, axonal growth/guidance, neuronal apoptosis, patterning, and injury protection in the mammalian nervous system [5–8]. Dysregulation in the TRK signaling leads to neural developmental disorders, various types of carcinomas, and also reported certain diseases of the nervous system [9].

The Tropomyosin receptor kinases (Trks) are of three subtypes: Trk-A, Trk-B, Trk-C, and all three possess high similarity in terms of their structure [10]. The percentage of structural protein sequence accounts as follows: for Trk-A and Trk-B is 49%, for Trk-A and Trk-C it is 51%, and for Trk-B and Trk-C it is 55%. For all three Trks, the kinase domain sequence percent homology is 71% [11]. Despite having structural similarities, all three Trks show distinct roles in central nervous system (CNS) development. Trk-A possesses an important role in the normal sympathetic neuron development, Trk-B is important and required for proper development of sensory neurons, and Trk-C is important during early stages of sympathetic neuron development [12, 13]. Four ligands, namely neurotrophins: Neurotrophin-3 (NT-3), Neurotrophin-4 (NT-4), nerve growth factor (NGF) and brain-derived neurotrophic factor (BDNF) activate the Trk receptors [14–16].

After a TRK B scintillation proximity assay, Choi and colleagues conducted in 2008, screening of various compounds resulted in a potent TRK B inhibitor (IC_{50} Z 83 nmol/L) with a novel benzonitrile substituted imidazopyridine core [17]. Series of optimizations to that compound lead to the discovery of the most promising 2 FDA approved TRK Inhibitors, i.e., Larotrectinib and Entrectinib. Structurally, named as N-[5-[(3,5-difluorophenyl)methyl]-1H-indazol-3-yl]-4-(4-methylpiperazin-1-yl)-2-(oxan-4-ylamino)benzamide belongs to a class of phenylpiperazines with 3-indazolyl group in its core skeleton (Figure 15.1).

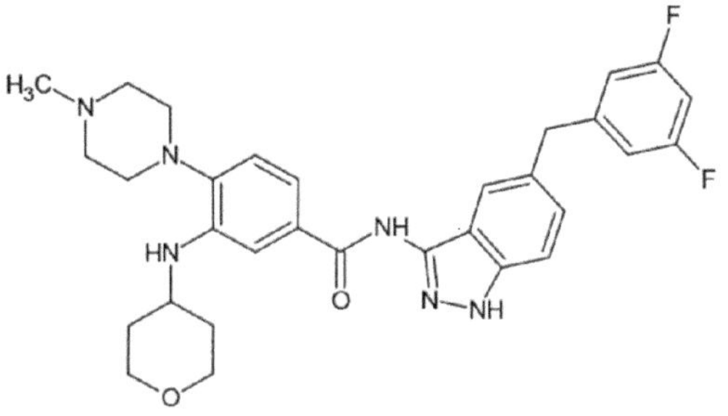

FIGURE 15.1 Structure of entrectinib.

It is approved as the third "tissue agnostic" cancer drug. "Tissue agnostic" means that the tumors are with a specific genetic feature in place of a specific tumor type. It acts as a monotherapy against patients with advanced non-small cell lung cancer (NSCLC) and patients suffering from solid tumors expressing NTRK gene fusions with a locally advanced disease or may be metastatic or where surgical procedure would result in severe morbidity [18]. The fusion of one of the three *NTRK* genes (Neurotropic tyrosine receptor kinase) with another gene as a result of specific genetic alteration harbored in solid tumors leads to the production of TRK fusion proteins. In patients with such kind of alteration, entrectinib inhibits the TRK activity and other proteins that can promote cancer growth by acting as a targeted therapy. In the case of adults, NSCLC with a gene fusion involving the *ROS1* gene (a gene that is structurally similar to the anaplastic lymphoma kinase (ALK) protein). Entrectinib also restricts the activity of ROS1 proteins [19].

It is the first-generation TRK inhibitor showing activity on all 3 TRK A, TRK B, TRK C inhibitor subtypes and was granted a breakthrough designation by the US FDA for the treatment of advanced or recurrent TRK fusion-positive cancers in May of 2017 and consequently in Japan as well in 2019 [20]. The presence of 2 nitrogen atoms in indazole ring and the –NH– in benzamide group are highly important for protein interaction of (PDB ID: 5KVT) at Met592 and Glu590 sites [21].

15.2 MATERIALS AND METHODS

15.2.1 DESIGN OF BIOISOSTERES OF ENTRECTINIB

The MolOpt tool is a user-friendly and robust web server for *in-silico* drug design using bioisosteric transformation rules derived from data mining, deep generative models, and similarity comparison to identify new bioisosteric transformation ideas. Here, the design of novel *in-silico* analogs was done by replacing the phenyl moiety of benzamide group to produce less toxic and more potent analogs with fewer side effects than the standard drug (Entrectinib) by bioisosteric approach using MolOpt software, and further modification of the bioisosteres was done based on the ADMET results. This service is freely accessible to the public at http://xundrug.cn/molopt [22]. The structures of analogs were drawn using ACD/ChemSketch 2021.1.1 (www.acdlabs.com).

15.2.2 PHARMACOKINETIC (ADME) AND TOXICITY PROPERTIES PREDICTION

ADMETLab 2.0 is an enhanced version online web tool for systematical evaluation of medicinal chemistry friendliness, pharmacokinetic (ADME) and a few physicochemical properties of a drug molecule. The software predicts 88 related characteristics belonging to 7 different categories. The software is available freely, without registration, at https://admetmesh.scbdd.com/ [23].

15.3 RESULTS AND DISCUSSION

15.3.1 BIOISOSTERES OF ENTRECTINIB

On the basis of existing scientific evidence, a series of entrectinib analogs were designed based on bioisosteric replacement using the MolOpt tool, and further modifications led to a few more analogs as per the ADMET results.

Furthermore, these analogs were investigated for ADMET prediction by an online tool – ADMETLab 2.0. Table 15.1 indicates the predicted molecular properties like molecular weight (MW), total count of hydrogen bond acceptors (nHA), total count of hydrogen bond donors (nHD), and the LogP value at different pH of the newly designed analogs. It was found to fulfill the criteria for Lipinski rule of five, which was used in the prediction of oral bioavailability of potential lead molecules as well.

TABLE 15.1 Bioisosteric Analogs of Entrectinib and Their Predicted Molecular Properties

Entry No.	Chemical Structure	MW	nHA	nHD	TPSA	LogS	LogD	LogP
(04) 30	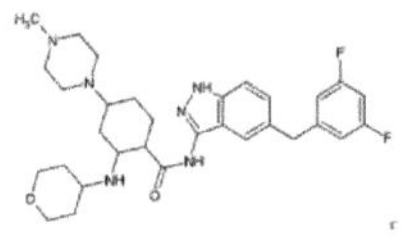	566.32	8	3	88.75	–2.192	3.059	2.963
(04) 49	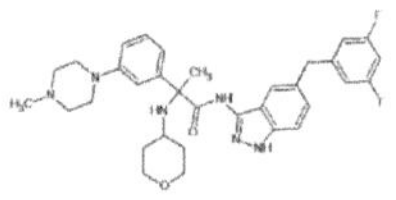	588.3	8	3	88.75	–3.953	4.17	4.576
1	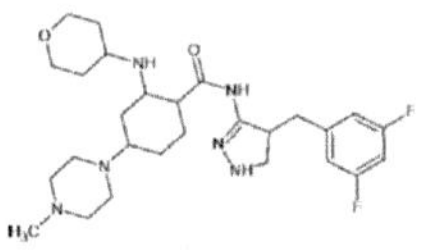	518.32	8	3	81.23	–1.104	1.705	1.274
3	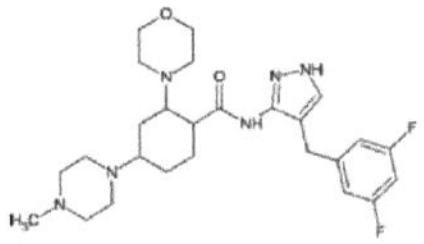	502.29	8	2	79.96	–2.149	1.946	1.906
5	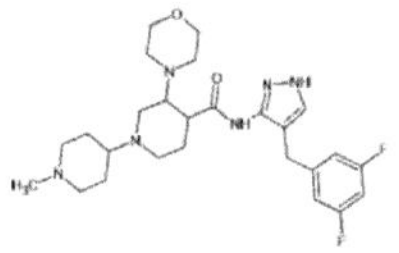	502.29	8	2	79.96	–1.828	1.835	1.789
7	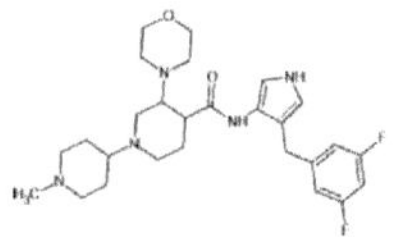	501.29	7	2	63.84	–2.189	1.925	2.337
9	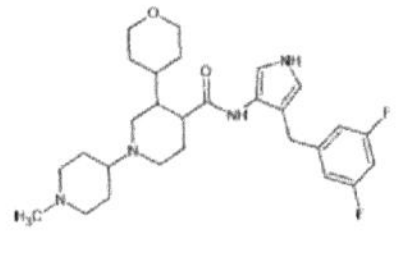	500.3	6	2	60.6	–3.365	2.676	3.295

TABLE 15.1 *(Continued)*

Entry No.	Chemical Structure	MW	nHA	nHD	TPSA	LogS	LogD	LogP
11	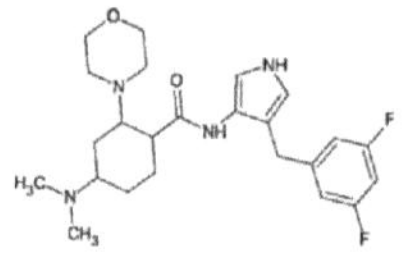	446.25	6	2	60.6	–2.801	2.369	2.648
13	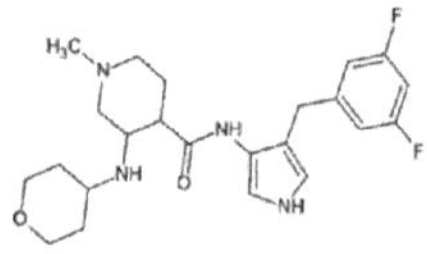	432.23	6	3	69.39	–1.615	1.982	2.044
15	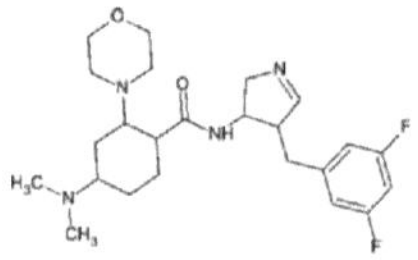	448.26	6	1	57.17	–2.205	2.562	2.36
19	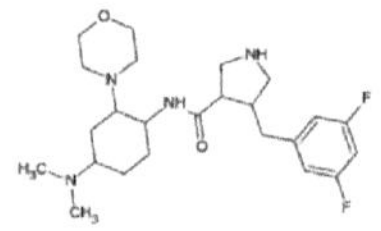	450.28	6	2	56.84	–1.596	1.659	1.445
Entrectinib	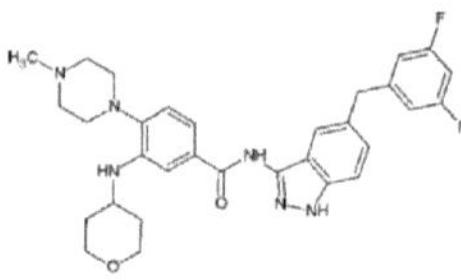	560.27	8	3	88.75	–5.735	3.826	4.872

Abbreviations: MW: Molecular weight; nHA: total count of hydrogen bond acceptors; nHD: total count of hydrogen bond donors; TPSA: topological polar surface area; logS: the logarithm of aqueous solubility value; logD: logarithm of n-octanol/water distribution coefficients at physiological pH of 7.4 (logD7.4); and logP: logarithm of n-octanol/water partition coefficient.

15.3.2 PREDICTION OF PHARMACOKINETIC (ADME) AND TOXICITY PROPERTIES OF ENTRECTINIB

The ADMELab 2.0 software was used to predict various pharmacokinetic (ADME) and toxicity parameters. Medicinal chemistry related properties of the analogs are illustrated in Table 15.2. Compounds 11, 15, and 19 have excellent QED values (a measure of drug likeness (DL)), i.e., > 0.67, showing that the above analogs are highly attractive. The synthetic acceptability score of all analogs was found to be excellent, i.e., < 5, indicating that the analogs are easy to synthesize. All analogs were following Lipinski and Pfizer rules, whereas 33% of all analogs were following Golden Triangle rules, excluding Entrectinib.

TABLE 15.2 MedicinalChemistry Properties of the Designed Entrectinib Analogs

Entry No.	QED	SA Score	Fsp3	MCE-18	Lipinski	Pfizer	GSK	Golden Triangle
(04) 30	0.426	4.374	0.548	116	Accepted	Accepted	Rejected	Rejected
(04) 49	0.301	3.825	0.394	115.304	Accepted	Accepted	Rejected	Rejected
1	0.529	4.584	0.704	100.957	Accepted	Accepted	Rejected	Rejected
3	0.652	4.303	0.615	99	Accepted	Accepted	Rejected	Rejected
5	0.652	4.172	0.615	99	Accepted	Accepted	Rejected	Rejected
7	0.638	3.812	0.593	98.581	Accepted	Accepted	Rejected	Rejected
9	0.622	3.846	0.607	99	Accepted	Rejected	Rejected	Rejected
11	0.715	3.927	0.542	81.432	Accepted	Accepted	Rejected	Accepted
13	0.657	3.768	0.522	78.571	Accepted	Accepted	Rejected	Accepted
15	0.725	4.454	0.667	83.95	Accepted	Accepted	Rejected	Accepted
19	0.690	4.179	0.708	84.707	Accepted	Accepted	Rejected	Accepted
Entrectinib	0.324	3.235	0.355	77.714	Accepted	Accepted	Rejected	Rejected

Abbreviations: QED: Related to drug-likeness; SA Score: synthetic accessibility score; Fsp3: total count of sp3 hybridized carbons; and MCE-18: medicinal chemistry evolution in 2018.

Various absorption properties (Caco-2, MDCK, HIA) and distribution properties (PPB, VDss, BBB, and Fu percentage) are illustrated under Table 15.3. Only 3 analogs (3, 11, and 15) show excellent Caco-2 permeability and the rest of the compounds show comparatively better permeability than Entrectinib. All analogs possess excellent MDCK permeability and HIA values. An excellent HIA value indicates better oral absorption, consequently assuring that the analog may be orally effective. All analogs possess a bit more BBB penetration than Entrectinib, thus facilitating entry into the brain. All the analogs except Entrectinib possess a predicted PPB value of < 90% and are considered excellent. All the analogs possess an excellent volume of distribution (within 0.04–20 L/kg). Metabolism (CYP1A2-inh and -sub, CYP2D6-inh and -sub) and excretion (CL and $T_{1/2}$) properties are also enlisted. Around 98% of the designed analogs have a CL value greater than 5, and 90% have a $T_{1/2}$ value within the range of 0–0.3, thereby displaying excellent clearance in comparison to Entrectinib.

TABLE 15.3 ADME Properties of the Designed Entrectinib Analogs

E. No.	Caco-2	MDCK	HIA	BBB	PPB (%)	VD	Fu (%)	C1A2 (i)	C1A2 (s)	C2D6 (i)	C2D6 (s)	CL	T½
(04) 30	–5.535	Ex.	0.079	0.85	41.81	1.516	34.25	—	—	–	++	5.075	0.074
(04) 49	–5.727	Ex.	0.226	0.894	89.05	2.359	5.65	—	–	++	++	4.917	0.071
1	–5.406	Ex.	0.015	0.643	32.60	0.877	64.74	—	—	—	++	10.85	0.141
3	–4.868	Ex.	0.017	0.928	16.61	1.11	57.84	—	—	+	++	8.531	0.194
5	–5.456	Ex.	0.023	0.892	15.31	1.152	61.45	—	—	–	++	7.511	0.195
7	–5.336	Ex.	0.007	0.956	40.29	1.376	54.66	—	—	+	++	10.63	0.284
9	–5.332	Ex.	0.004	0.803	44.66	2.021	47.37	—	—	++	++	11.54	0.114
11	–5.133	Ex.	0.004	0.954	39.85	1.345	55.55	—	–	++	++	11.94	0.512
13	–5.47	Ex.	0.009	0.89	23.73	1.425	66.24	—	–	++	+	8.905	0.459
15	–4.856	Ex.	0.004	0.935	42.82	1.47	55.39	—	—	++	++	11.15	0.082
19	–5.595	Ex.	0.005	0.766	25.40	1.189	69.86	—	—	–	++	11.26	0.16
Ent.	–5.596	Ex.	0.231	0.509	96.40	2.56	3.02	–	+	++	+++	4.15	0.06

Abbreviations: E. No.: Entry number; Caco-2: the human colon adenocarcinoma cell lines; MDCK: Madin-Darby Canine kidney cells (an *in vitro* model for screening of permeability) and >2 × 10^{-6} cm/s indicates excellent; HIA: human intestinal absorption; BBB: blood-brain barrier penetration; PPB: percentage plasma protein binding; VD: volume of distribution; Fu: the fraction of unbound drug in plasma; human cytochrome P450 isozymes; CYP1A2: C-1A2; CYP2D6: C-2D6; (–) indicates not an inhibitor/substrate; (+) indicates an inhibitor/substrate; i: inhibitor; s: substrate; CL: clearance of the drug; T ½: the half-life of a drug; Ex.: excellent; and Ent: entrectinib.

Toxicity properties (hERG, H-HT, DILI, AMES, and NR-AR) are illustrated in Table 15.4. The regulation of the exchange of cardiac action and resting potential is due to the human ether-a-go-go related gene (hERG). Only 3, 5, and 19 analogs possessed excellent activity and less hERG blockade. The human hepatotoxicity (H-HT) value of all analogs is less than that of Entrectinib, but analogs 7, 11, 13, 15, and 19 possess medium toxicity as they are in the 0.4–0.7 range. Almost 90% of analogs possess lesser DILI values and fall in the range of excellent category than Entrectinib. All analogs, except Entrectinib itself, have no Ames toxicity as they fall under the excellent category (< 0.3). Except for 1, 15, and Entrectinib, all other analogs have less carcinogenicity. All compounds, including Entrectinib, possess respiratory toxicity. Although all values for NR-AR and NR-AR-LBD receptors are excellent, the analogs have no activity against the androgen receptor (AR). Finally, out of the newly designed Entrectinib analogs, 11, 15, and 19 indicate promising results and may act as good leads based on the QED values (> 0.67) and low toxicity profile (hERG, H-HT, and DILI) compared to that of the existing marketed drug of Entrectinib with better efficacy. This *in-silico* study may act as a way forward for drug designing and development of newer and potent TRK Inhibitors.

TABLE 15.4 Toxicity Profile of the Designed Entrectinib Analogs

Entry No.	hERG	H-HT	DILI	Ames	Carc.	Resp.	NR-AR	NR-AR-LBD
(04) 30	0.826	0.919	0.176	0.038	0.132	0.983	0.067	0.003
(04) 49	0.694	0.956	0.185	0.055	0.061	0.819	0.075	0.003
1	0.384	0.952	0.027	0.037	0.423	0.912	0.001	0.001
3	0.302	0.881	0.042	0.026	0.053	0.963	0.058	0.003
5	0.265	0.879	0.045	0.029	0.066	0.978	0.157	0.003
7	0.436	0.431	0.051	0.016	0.03	0.973	0.176	0.002
9	0.835	0.406	0.316	0.009	0.025	0.969	0.143	0.002
11	0.515	0.44	0.139	0.013	0.038	0.923	0.002	0.002
13	0.577	0.645	0.325	0.015	0.03	0.96	0.033	0.002
15	0.409	0.479	0.025	0.018	0.534	0.928	0.002	0.001
19	0.241	0.379	0.022	0.013	0.121	0.919	0.001	0.001
Entrectinib	0.838	0.989	0.971	0.4	0.386	0.961	0.347	0.012

Abbreviations: hERG: hERG blockers; H-HT: the human hepatotoxicity; DILI: drug-induced liver injury; Ames: test for mutagenicity; Carc.: carcinogenicity; Resp.: respiratory toxicity; NR-AR: a nuclear hormone receptor-androgen receptor; and NR-AR-LBD: whether the drug molecule attaches to the ligand binding domain (LBD) of androgen receptor.

15.4 CONCLUSION

Entrectinib analogs were designed by bioisosteric approach based on the published scientific evidence. Their ADMET properties were predicted by various computational tools. The objective of the *in-silico* design was to develop less toxic and more potent analogs with fewer side effects than the standard drug (Entrectinib) and further modifications were performed based on the ADMET results. Analogs 11, 15, and 19 are found prominent based on their least toxicity levels (DILI score < 0.3) and exhibiting potential drug-likeness (QED > 0.67). At the same time, all designed analogs were found to show better results in suitable parameters than the standard drug. Therefore, these newly designed analogs may have potential to be used as lead compounds, if further found to deliver promising and reliable experimental results as potent TRK inhibitors.

KEYWORDS

- **ADMETLab 2.0**
- **bioisosteric approach**
- **entrectinib analogs**
- **MolOpt**
- **pharmacokinetic properties**
- **toxicity properties**
- **Trk inhibitor**

REFERENCES

1. Siegel, R. L., Miller, K. D., Fuchs, H. E., & Jemal, A., (2021). Cancer statistics, 2021. *CA: A Cancer Journal for Clinicians*, *71*, 7–33. doi: https://doi.org/10.3322/caac.21654.
2. Gagic, Z., Ruzic, D., Djokovic, N., Djikic, T., & Nikolicet, K., (2020). In-silico methods for design of kinase inhibitors as anti-cancer agents. *Frontiers in Chemistry*, *7*, 1–25.
3. Grossman, M., & Adler, E., (2021). *Protein Kinase Inhibitors – Selectivity or Toxicity?*IntechOpen. doi: Http://dx.doi.org/10.5772/intechopen.98640.
4. Huang, L., Jiang, S., & Shi, Y., (2020). Tyrosine kinase inhibitors for solid tumors in the past 20 years (2001–2020). *Journal of Hematology and Oncology*, *13*, 1–23. doi: Https://doi.org/10.1186/s13045-020-00977-0.
5. Chan, J. R., Cosgaya, J. M., Wu, Y. J., & Shooter, E. M., (2001). Neurotrophins are key mediators of the myelination program in the peripheral nervous system. *The Proceedings of the National Academy of Sciences, U.S.A.*, *98*, 14661–14668.
6. Hempstead, B. L., & Salzer, J. L., (2002). Neurobiology. A glial spin on neurotrophins. *Science*, *298*, 1184–1186.
7. Patapoutian, A., & Reichardt, L. F., (2001). Trk receptors: Mediators of neurotrophin action. *Current Opinion in Neurobiology*, *11*, 272–280.
8. Sofroniew, M. V., Howe, C. L., & Mobley, W. C., (2001). Nerve growth factor signaling, neuroprotection, and neural repair. *Annual Review of Neuroscience*, *24*, 1217–1281.
9. Stoleru, B., Popescu, A. M., Tache, D. E., Neamtu, O. M., Emami, G., Tataranu, L. G., Buteica, A. S., et al., (2013). Tropomyosin-receptor-kinases signaling in the nervous system. *MEDICA: A Journal of Clinical Medicine*, *8*, 43–48.
10. Nakagawara, A., (2001). Trk receptor tyrosine kinases: A bridge between cancer and neural development. *Cancer Letters*, *169*, 107–114.
11. Nakagawara, A., Liu, X. G., Ikegaki, N., White, P. S., Yamashiro, D. J., Nycum, L. M., Biegel, J. A., & Brodeur, G. M., (1995). Cloning and chromosomal localization of the human TRK-B tyrosine kinase receptor gene (NTRK2). *Genomics*, *25*, 538–546.
12. Fagan, A. M., Zhang, H., Landis, S., Smeyne, R. J., Santiago, I. S., & Barbacid, M., (1996). TrkA, but not TrkC, receptors are essential for survival of sympathetic neurons *in vivo*. *The Journal of Neuroscience*, *16*, 6208–6218.

13. Barbacid, M., (1995). Structural and functional properties of the TRK family of neurotrophin receptors. *Annals of the New York Academy of Sciences*, *766*, 442–458.
14. Kaplan, D. R., & Miller, F. D., (2000). Neurotrophin signal transduction in the nervous system. *Current Opinion in Neurobiology*, *10*, 381–391.
15. Huang, E. J., & Reichardt, L. F., (2001). Neurotrophins: Roles in neuronal development and function. *Annual Review of Neuroscience*, *24*, 677–736.
16. Bibel, M., & Barde, Y. A., (2000). Neurotrophins: Key regulators of cell fate and cell shape in the vertebrate nervous system. *Genes & Development*, *14*, 2919–2937.
17. Choi, H. S., Rucker, P. V., Wang, Z., Fan, Y., Albaugh, P., Chopiuk, G., Gessier, F., et al., (2015). (R)-2-Phenylpyrrolidine substituted imidazopyridazines: A new class of potent and selective pan-TRK inhibitors. *ACS Medicinal Chemistry Letters*, *6*, 562–567.
18. Imran, M., Asdaq, S. M. B., Khan, S. A., Unnikrishnan, M. D., Alamri, A. S., Alsanie, W. F., Alhomrani, M., et al., (2021). Innovations and patent trends in the development of USFDA approved protein kinase inhibitors in the last two decades. *Pharmaceuticals*, *14*, 710. https://doi.org/10.3390/ph14080710.
19. National Cancer Institute, (2019). *FDA Approves Entrectinib for Tumors with NTRK Fusions*. Cancer Current Blogs.https://www.cancer.gov/news-events/cancer-currents-blog/2019/fda-entrectinib-nTRK-fusion (accessed on 21 December 2023).
20. Drilon, A., (2019). TRK inhibitors in TRK fusion-positive cancers. *Annals of Oncology: Official Journal of the European Society for Medical Oncology*, *30*, viii23–viii30. https://doi.org/10.1093/annonc/mdz282.
21. Jiang, T., Wang, G., Liu, Y., Feng, L., Wang, M., Liu, J., Chen, Y., & Ouyang, L., (2021). Development of small-molecule tropomyosin receptor kinase (TRK) inhibitors for *NTRK* fusion cancers. *Acta Pharmaceutica Sinica*, *11*(2), 355–372. https://doi.org/10.1016/j.apsb.2020.05.004.
22. Shan, J., & Ji, C., (2020). MolOpt: A web server for drug design using bioisosteric transformation. *Current Computer-Aided Drug Design*, *16*, 460–466.
23. Dong, J., Wang, N. N., Yao, Z. J., Zhang, L., Cheng, Y., Ouyang, D., Lu, A. P., & Cao, D. S., (2018). ADMETlab: A platform for systematic ADMET evaluation based on a comprehensively collected ADMET database. *Journal of Cheminformatics*, *10*, 1–11.

CHAPTER 16

IMPACT OF TALENT MANAGEMENT STRATEGIES ON RETAINING THE TALENTED EMPLOYEES OF THE PHARMACEUTICAL INDUSTRY.

AMBARISH GHOSH,[1] MANOJ VERGHESE,[2] and J. H. VYAS[3]

[1]*Research Scholar, CSVTU, Bhilai, Chhattisgarh, India*

[2]*Professor, RCET, Bhilai, Chhattisgarh, India*

[3]*Professor, RIT, Raipur, Chhattisgarh, India*

ABSTRACT

The organization's success depends on the human resources of that organization, but the pharmaceutical industries of India are facing the problem of retaining talented employees. The challenge for managers today is to maintain employee engagement and ensure that they remain in the organization for a longer period of time, which would be beneficial for both the employee and the employer. The manager of the pharmaceutical organization must be aware of the current recruitment and selection procedures and retention strategies of their organization. The purpose of this research is to discover whether there is any impact of talent management strategies on retaining talented employees in the pharmaceutical organization. In this research chapter, a descriptive study was conducted within different pharmaceutical industries, and in this study, five companies were taken. The target population consisted of medical representatives and executives. A structured questionnaire was distributed, and primary data was collected using a simple random sampling method. Analysis was done using the SPSS version 21.0 package. The data

Computational Optimization, Modeling, and Simulation for Engineering Applications.
Anupam Shukla, Sourabh Rungta, Mohan Awasthy, and Rakesh Himte (Eds.)

was evaluated and summarized in frequency tables in order to determine the number of responses and variables with each determination. To examine the strength and association of different factors, the study used correlation and regression analysis at the 5% significant level.

From the total analysis, it was clear that talent acquisition had a positive and maximum influence (Pearson's correlation coefficient = 0.877) on retaining the talented employee. Additionally, talent management, compensation strategies, and succession planning are positively correlated with retaining the talented employee. From the findings of the study, it is clear that HR management in pharmaceutical industries has very little concern on career planning and succession planning strategies adopted, and in most organizations, employees are not willing to stay with the organization for a longer period of time. The study also recommends pharmaceutical companies to conduct induction programs nicely and provide the best knowledge to newly appointed employees, which helps in retaining talented employees.

- **Broad Objective:** To analyze the impact of talent management strategies and to find out how talented employees can be retained in a pharmaceutical organization for a longer period of time.
- **Particular Objective:**
  - To analyze the impact of strategies of talent acquisition on retaining the talented employee in the pharmaceutical organization.
  - To analyze the impact of talent development strategies on retaining talented employees in the pharmaceutical organization.
  - To analyze the impact of compensation strategies adopted on retaining talented employees in the pharmaceutical organization.
  - To discover the strategies for succession planning and retaining talented employees in the pharmaceutical organization.
- **Conceptual Framework:**

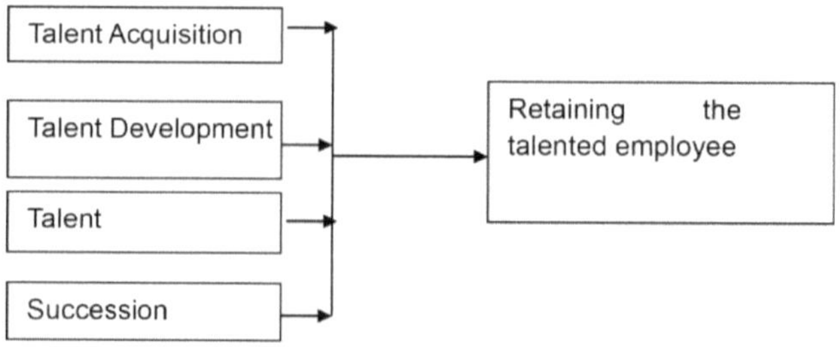

16.1 RESEARCH METHODOLOGY

The purpose of research is to find answers to questions and to gain new knowledge about the subject of study. Because research is the major technique utilized in practically all fields of organizational research to widen and narrowly arrive at precise findings and recommendations, it aids in expanding the frontiers of knowledge. To perform the study successfully, it is a systematic approach in which we find research design, target population, instruments for data collection, and techniques of data analysis [1–14].

16.1.1 RESEARCH DESIGN

A research design is the blueprint of the total analysis, which shows how data is to be collected and analysis was done and to find out its purpose with economy in procedure. In this chapter, descriptive research design was used to collect data and analyze it.

16.1.2 TARGET POPULATION

In this research, the researcher had targeted five pharmaceutical companies: Aristo Pharmaceuticals, Ajanta Pharmaceuticals, Palsons Drugs and Chemical Industries, Torrent Pharmaceuticals, and USV Pharmaceuticals Ltd. The targeted population was managers, medical representatives, and the sample size is 200.

16.1.3 SAMPLING FRAME

The sampling process used in this research was the stratified sampling method, and employees of the five pharmaceutical companies were clustered into five strata.

16.1.4 INSTRUMENTS USED IN RESEARCH

Questionnaire was the main tool for the study and survey was done to collect data from 200 sales force of five pharmaceutical companies. Stratified random sampling process was used in the total process. 10 strata were formed from clusters of employees. The stratum was based on the company's sales force and the management teams of the company, either a manager or a medical representative.

16.1.5 AGE AND GENDER OF THE RESPONDENTS

The plurality of respondents, 44.7%, were between the ages of 36 and 40. 20% were between the ages of 31 and 35. 14.4% were between the ages of 41 and 45. 8.9% were between the ages of 25 and 30. 6.5% were over 45, and just 5.6% were under the age of 25. According to the study's analysis, data was obtained from people of all ages and majority, i.e., 89.7%, were male and only a few, 10.3%, of the respondents were female.

16.1.6 EDUCATIONAL BACKGROUND OF THE RESPONDENTS

The study shows that the education of the respondents, majority (39.6%) were graduates (34.0%), have professional degrees, i.e., MBA (20.3%), were post-graduates in science and only 6.1 have some other degree. From this study it is quite clear that the respondents are well qualified and therefore they can understand the intention of the study and the information they provide for quality research.

16.1.7 EMPLOYEE WORKING DURATION IN YEARS

In this study, the findings on the age of the respondents were that the majority, i.e., 48.2%, had worked for between 0 and 5 years, 19.7% for between 5 and 10 years, while 16.5% for 10-15 years. Only a few, i.e., 3.6%, had worked for over 20 years. The data shows that the respondents can understand the working and strategies of the organization; therefore, they can provide a good view of the research purpose.

16.2 TALENT ACQUISITION

From the whole study, the author finds many talent acquisition strategies, which are effective to retain talented employee retention in the pharmaceutical industry.

16.2.1 RETAINING THE TALENTED EMPLOYEE BY ACQUIRING THE RIGHT TALENT

The data of the study reveals that 68.0% of the employees believe that various talent acquisition strategies have an impact on employee retention in the pharmaceutical industry, while 32.0% believe that there is no impact of talent acquisition strategies on retaining the talented employees in a pharmaceutical company.

16.2.2 STRATEGIES FOR ATTRACTING TALENTED EMPLOYEE

The data was collected and studied, which reveals that the majority of the employees, i.e., 34.7%, were attracted by various incentive and compensation plans. 32.3% of employees are attracted by various training programs, 19.4% of employees are attracted by a well-developed career path, while a few, i.e., 5.6%, were attracted by flexible working hours. From the analysis, it is clear that the majority of the respondents say various incentive and compensation plans are necessary to attract new talents. This reveals the implication that various incentive and compensation plans make the employees feel that the companies are concerned for them, and the pharmaceutical organization can use this strategy to retain their talented employees.

16.2.3 TALENT DEVELOPMENT

The study reveals that the majority of pharmaceutical organizations organize quarterly meetings and, in those meetings, organize many workshops and management development programs to support and retain talented employees. The company also organizes new leader workshops to assist the sales force in making them well-equipped to take on new responsibilities.

Majority of the employees believe that the organization has started guidance and leadership development programs that can develop talent management and help in retaining talented employees. The study shows that pharmaceutical companies have to implement leadership. The results show that most pharmaceutical companies have started implementing leadership programs which can develop talent management and retain the best talented employees in the organization.

16.2.4 STRATEGIES FOR COMPENSATION ACCORDING TO TALENT

From the data, it can be concluded that the pharmaceutical organization has adopted compensation strategies to improve talent management that can retain talented employees in the pharmaceutical company.

16.2.5 STRATEGIES TO IMPROVE TALENT MANAGEMENT AND RETAINING THE TALENTED EMPLOYEE

From the data findings, the researcher's findings revealed that most of the respondents think that pay should be according to performance and budget compensation. The Pharmaceutical organization should make a benchmark according to competitors and make a pay range which is suitable. The study revealed that 89.2% of the respondents fixed the benchmark of the organization, which is parallel to the competitors, so that the organization may come up with suitable pay. 10.8% of the respondents don't agree with the opinion. According to this survey, 12.0% of respondents differ with the rewards provided by the organization and are corresponding with the effort, and 5.0% are not sure if the rewards are commensurate with their efforts. According to 60.0% of respondents, the rewards provided by this organization are commensurate with the work they put into it, whereas 20.0% strongly agreed.

16.3 SUCCESSION PLANNING

Within the pharmaceutical industry, the study determines that if succession planning strategies had an effect on retaining the talented employee in a pharmaceutical organization.

16.3.1 USING SUCCESSION PLANNING STRATEGIES TO RETAIN EMPLOYEES

Employees believe that succession planning is adopted in most pharmaceutical companies. The employees in the organization always try their level best for their performance and do better according to their capacity, capabilities, and their ideas. The employees always try their level best for performance, which is beneficial to them only.

16.4 EMPLOYEE RETENTION

The study reveals that every pharmaceutical company organizes an induction/training program for their new employees, so that they can understand their duties and responsibilities and how they can perform better in the organization. The manager also assists the newly appointed employee when they are in the field.

16.5 REGRESSION ANALYSIS

In this study, we have applied multiple regression analysis to find the impact of talent management strategies on retaining talented employees in a selected pharmaceutical organization. The Statistical Package for the Social Sciences (SPSS) was used to analyze the data and compute the results using multiple regression analysis (Table 16.1).

TABLE 16.1 Model Summary on Retaining the Talented Employee

Model	R	R Square	Adjusted R Square	St. Error of Estimation
1	0.898	0.807	0.234	0.277

The coefficients of the determinants explain how changes in independent variables can make changes in dependent variables. In this study, the dependent variable is retaining the talented employee and the independent variables which we have taken are talent acquisition, talent development, compensation, and succession planning. So, from this study, it is clear that the value of R square is 80.7%. Therefore, the required factors were considered for retaining talent, which is the result of R square (Table 16.2).

TABLE 16.2 ANOVA on Retaining the Talented Employee

Model	Sum of Squares	df	Mean Square	F	Sig.
Regression	1.34	10	0.25	53.11	0.00
Residual	99.90	57	1.26		
Total	101.24	67			

16.6 ANOVA

Talent acquisition, talent development, compensation strategies, and succession planning are the predictors (constants). Retaining the talented employees is the variable.

The significance value is 0.00 which is less than 0.05 thus the model is statistically significant. The F critical at 5% significance level is 7.9. Hence, the calculated value of F is larger than the critical value of F (value = 53.1233). We can therefore conclude that the model is significant in totality.

As the result 0.00 is less than 0.05, therefore we can say that model which we have made is statistically significant in. As in our result, F is larger than the critical value of F (value = 53.1233) at a 5% significance level.

16.7 CONCLUSION

The study concluded that as a talent attraction strategy, it suggests that organizations should focus on financial benefits, which will surely have a favorable outcome and attract more talented employees.

From this study, we can also conclude that the pharmaceutical business enterprise must guarantee that they come with board members who are well-prepared in all aspects to meet the top management and who can perform.

This research chapter also concluded that talent management and planning workforce are followed by pharmaceutical organizations as skills improvement techniques to improve talent management and to retain talented employees. The organization should emphasize how the talent can be used effectively to the best use, and the workforce planning should be such that it guarantees all of the respective skills for their advantaged use. Every worker should receive a deal according to their capabilities.

The examiner concluded that the business enterprise has a well-evolved onboarding program for brand new personnel. This means that for each new worker, they may be comfortable considering they the onboard program.

16.8 RECOMMENDATION OF THE STUDY

Pharmaceutical organization needs to develop a good onboarding process for new hires to enable onboarding of relevant competent employees during the transition. Organizations need to improve their employee learning needs analysis. Competitive salaries should be there, and remuneration is based on business success, so we don't need to set up a system to guarantee that your staff are aware of this system and are fairly rewarded. Organizations need to be able to provide benefits in a variety of ways, as they can provide benefits to their employees for motivation and can cover and protect them in specific aspects.

KEYWORDS

- **pharmaceutical organization**
- **regression analysis**
- **retaining talented employee**
- **succession planning**
- **talent acquisition**
- **talent compensation**
- **talent management strategies**

REFERENCES

1. Abhilasha, S., & Jyoti, S., (2015). Strategies for talent management, a study of select organizations in the UAE. *International Journal of Organizational Analysis*.
2. Bhatnagar, J., (2007). Predictors of organizational commitment in India: Strategic HR roles, organizational learning capability and psychological empowerment. *International Journal of Human Resource Management, Forthcoming*.
3. Brittain, S., (2006). *Talent Management Selection and Development Review, 22*(3).
4. Collings, D. G., & Mellahi, K., (2009). *Strategic Talent Management*, A review and research agenda. *Human Resource Management Review, 19*(4), 304–313.
5. Dana, E., Josef, M., & Iveta, U., (2013). *Integrated Talent Management*. Challenge and future for organizations in the Visegrad Countries Published in the Czech Republic.
6. Hilligoss, B., & Rieh, S. Y., (2008). *Developing a Unifying Framework of Credibility Assessment, Construct, Heuristics, and Interaction in Context, 21*, 212–215.
7. Abadi, M., Khadem, S., & Sadeghnia, A., (2013). Talent management and maintaining talented human resource: A case study in Kashan university of medical sciences. *European Online Journal of Natural and Social Sciences, 2*(3), 3446–3457.
8. AbulMohaimen, M., (2013). Talent management: Three new perceptions intended for managing and retaining talents in Bangladesh. *Asian Business Review, 3*(5), 7–15.
9. Ahammad, T., (2017). Personnel management to human resource management (HRM): How HRM functions. *Journal of Modern Accounting and Auditing, 13*(9), 412–420.
10. Ajaz, K., (2014). The changing dynamics of talent acquisition: An analysis of recruitment trends, marketing strategies, and emerging software and services. *Academ. Arena, 6*(4), 67–73.
11. Beheshtifar, M., Nasab, H., & Moghadam, M., (2012). Effective talent management: A vital strategy to organizational success. *International Journal of Academic Research in Business and Social Sciences, 2*(12), 227–234.
12. Bryan, L., Joyce, C., & Weiss, L., (2006). *Making a Market in Talent* (Vol. 2, No. 2, pp. 98–109). McKinsey Quarterly.

13. Canedo, J. C., Graen, G., & Grace, M., (2017). Let's make performance management work for new hires: They are the future. *Organizational Dynamics*, 1–5.
14. Cappelli, P., & Keller, J. R., (2014). Talent management: Conceptual approaches and practical challenges. *Annual Review of Organizational Psychology and Organizational Behavior* (Vol. 1, pp. 305–331).

CHAPTER 17

POWER LINE FAULT DETECTION

HARGOBIND PATEL, PARUL KHATRI, and SHRIKANT ZADE

Department of Computer Science and Engineering, OIST, Bhopal, Madhya Pradesh, India

ABSTRACT

Insulated overhead wires are increasingly being used in many regions throughout the globe due to their greater indestructible operating quality, end-of-phase contact, and distance between phases. However, ordinary insurance devices are not intended to consistently recognize a conductor's stage-to-ground issue and the increased chances of tree/tree branch mortality owing to incomplete release (PD) exercises on these instances. This is because an overcurrent is noticed on an open conductor. To address this issue, the Technical University of Ostrava (VSB) has developed an amazing meter that measures the voltage signal of the electric field as it moves through the protected air passages, with the goal of spotting unsafe PD activities above. In 2018, VSB shared a major portion of their recorded waveform data on Kaggle, the world's biggest computing cooperation platform, in order to investigate interesting instances of identification approaches for this application. With the availability of a huge dataset comprising routinely acquired high-repetition voltage signals, information-based analysis of reduction-related PD designs on a previously unheard-of scale becomes feasible. Because of the wide variety of PD instances and the foundation's restless limits, we devised a creative pulse shape delineation approach based on group processes that can strongly identify a group of pulses linked with the DP agent. Using these impulses as context, we build on the modest strengths

Computational Optimization, Modeling, and Simulation for Engineering Applications.
Anupam Shukla, Sourabh Rungta, Mohan Awasthy, and Rakesh Himte (Eds.)

and promote a deep learning model with unrivaled search performance for the driver challenges described in the early phase.

17.1 INTRODUCTION

Artificial intelligence (AI) is likely to be the most active issue of the decade. It has seen shaky development and needs to expand into practically every industry (design, metering, and control, biomedicine, and autonomous vehicles, in a few contexts). This would allow for a more accurate, quicker, and convenient layout. AI is seeing an extraordinary turn of events as a subset of AI, with numerous extant varieties and known applications, particularly in the artificial neural network arena. Deep learning capabilities and demonstration of convolutional neural networks (CNNs) have researchers intrigued. As a consequence, AI-based systems and applications offer tremendous promise in a variety of electrical force design disciplines. The problem of electrically unbreakable quality of strength equipment is closely associated with work issues, surges, and the need for high voltage protective frames for varied loads, including powerful electric fields. As a result, an active diagnosis of the following physical impairment measures in the security framework is required. The opening of the electric field in the protective frame is a factor that initiates and creates various sorts of electric triggers. These indicate the discharge of internal volatiles into pits known as voids and beyond the protective framework. Midway release (PD) refers to circumstances in which there is no total safety failure, i.e., no instantaneous cathode crossing. The permanent electrical stress Pd has an impact on the equipment's dependability and longevity. Neural networks are used in a variety of applications; they all have the same aim of being able to intuitively choose highlights from enormous datasets and summarize their responses to scenarios not encountered throughout the learning curve [1, 2]. CNNs, which have mostly replaced laddered perceptron-based networks (MLPs), are now commonly employed in sign and image processing. Over the last three decades, it has been clear that neural networks may be efficiently used for PD design identification, diagnostics, and sensing applications [3–8]. Measurable administrators are often generated from the DP dispersion and used in the characterization approach to decrease the complexity of acknowledgment interactions [2–4, 5–8]. Concrete phase target PD reduction was used in early applications owing to computational complexities [8, 9]. PD Design Acknowledgement Work has been done in a variety of domains, such as stage or pulse magnitude propagation [13], pulse time waveforms [4–6], and PD pictures [1–3]. This method's real test is based on overlaid designs of different deformations that occur in high voltage electrical protection [2, 8, 9]. An example of employing a neural network for mid-release photos is described in this chapter, which uses CNNs with DP images to comprehend the growing phases of high voltage electrical protection.

17.2 LITERATURE REVIEW

The research study "Deep learning-based approach for detecting symmetric and asymmetric transmission line defects" focuses on using the ANN approach to analyze the phase currents of a system model created using Simulink/MATLAB. The model's output data is then inputted into the ANN, which is trained and tested using the Python programming language. The collected findings demonstrate that the performance of the ANN is excellent [1]. Crowter and colleagues have utilized a sophisticated vector space clustering method to employ neural networks for defect discovery. They estimate the stresses of hydraulic actuator circuits using neural networks [13]. Tahar Buthiba presented a study on fault detection in extra high voltage transmission lines for high-speed insurance. This study utilizes terminating line information and field networks of artificial neurons. The suggested Neural Issue Finder and Finder were created by combining information from the specified power grid model with information from the safety and strength frameworks. This allows for the simulation of different burnout conditions. Three-point detectors are suggested, followed by a comparative evaluation of the proposed reduction detector structures. The goal is to determine which reduction detector structure of the simulated neural network offers the best exposures [2]. Signal preparation is a crucial component of computer-based remote protection solutions. It allows for the swift detection of problems and the restoration of power supply with minimal disturbance. By evaluating the coefficient of phase flow growth, it is possible to remove stage-specific evidence. An unbiased current smoothing coefficient test can be used to discover phase-to-phase ground issues. Reddy et al. developed a model based on a combination of impedance calculations and continuous wavelet transform methods. This model is used to identify the incidence of defects increasing with distance. MATLAB/Simulink is employed to create fault indications and validate computations.

The suggested fault location indicator seems to be promising based on replication findings [3]. Karthikeyan Kasinathan's The development of a theory aimed at the detection and grouping of force frame transmission line defects. A three-step augmented neural network strategy has been presented to deal with the issue of a very big collection of information with diverse defective circumstances. For discovery, the approach uses the Discrete Wavelet Transform, and for control, it employs two forms of self-coordinated and unaided adaptive resonance theory neural networks [4]. Shaban et al. [5] developed a power system framework devoid of an identifiable evidence pattern, identifying forms of transmission line failures utilizing data provided by wavelet evaluation of power system framework drifters. This chapter's study focuses on identifying observable evidence of force frame defects. Discrete wavelet transformation (DWT) The transitory destabilizing impact of event defects is investigated. Nature deposits and uses the most severe expansion coefficient, signal energy, and ratio of energy change of each kind of reconstructed original reduction to detect issue types [5]. Simulated neural networks were used for characterization and reduction. For each of the three stages, the fault zone measurement employs a feed-forward network with backpropagation computations. To accept the neural network's judgment, the neural network is checked on a neural network with a configurable number of layers covered and neurons per layer covered in each progression. The replication findings show that approaches based on simulated neural networks are successful in discovering transmission line defects and completing a tasteful repertoire [6]. Tayab and colleagues back-propagation (BP) neural network engineering is introduced as an alternate approach for minimizing space, sequence, and separation in a transmission line construction. The primary goal is to carry out a comprehensive teleprotection strategy for a transmission line structure. The remote protection job is separated into various neural networks to execute reduction discovery, demonstration of differentiating flaws (control) as a problem in distinct areas [7]. Duta et al. [8] proposed intelligent control to detect issues in transmission lines by employing neural networks customized by MATLAB programming to plan the required yield. The model classifier is a strong usage of artificial neural networks for learnability, synthesis, and uniform processing that may be utilized as a knowledgeable means for subsequent recognition. The Power Framework is used to investigate the disadvantages. The goal of this chapter is to put into action the aggregate strategy for distance protection, which is separated into several neural network regions. Faults are defined as one story to ground, two floors, and two floors to earth. Then comes a protection relay. The usage of neural network proliferation engineering for power transmission line framework should be conceivable [9]. In defect inference, fragile computational tools are seen as a significant extension of a quantitative model-based strategy for aging. When quantitative models are not readily available, an adequately modeled neural network may be utilized as a non-directly powerful framework model. Nonetheless, neural networks do not convey information effectively in the model.

Subjective performance or rule-based deduction strategies may be used to solve this underlying challenge. Srivastava et al. [10] discussed the advantages of several strategies for aggregating data into quantitative and subjective frameworks, as well as their relevant incentives for decreasing neural network analysis [10]. To detect the absence of transmission lines, Kesharwani et al. suggested a model based on MATLAB programming. The framework's performance is utilized to create simulated neural networks to detect transmission line issues. Simulated neural networks are used to do reduction search [11]. Desai demonstrates a novel method for identifying and resolving faults on transmission lines. The idea is to employ neural networks and wavelet transforms in tandem to choose a suitable solution to the challenge. The wavelet transform is made up of resilient, very quick, and exact numerical tools. For transient signals in transmission lines that have been employed in simulated neural networks to produce an alternation between the expected signal and the related signal, there are several instances. This may almost certainly be avoided by utilizing explicit calculations [12].

17.3 RESULT FROM ANALYSIS AND DISCUSSION

This chapter describes the use of CNNs for split-release image succession. In some constructions, partial release is often employed as one of the essential indicators of high voltage protection disruption. The test instance was developed under strong electric pressure, and the estimated outcomes were continually recorded within a predetermined time frame. The sequence of advancing Parkinson's disease (PD) photos from the prolonged mature trial was broken. Send the PD diagrams for the specific portions to the electrical safety maturation verification draw. The reported findings were created in Python using deep learning systems TensorFlow, Keras, and scikit-learn. In these cases, AI algorithms indicate that the data should be addressed and arranged in a two-dimensional cluster in a certain configuration ([samples, features]), where an example might be a PD picture and an element is an invariant. The acceptance accuracy of the model is 96.90%.

17.4 CONCLUSION

This chapter describes the use of a CNN to partially release photos in order to completely comprehend the maturation phases of high voltage electrical protection. The suggested model suggests a deterioration in electrical safety regulation. Step-by-step examples of PD photographs were discussed. The highlight of Applied Engineering was attempting to regulate the number of maps, the size of the convolution layers, and the bits as the hyper parameters' high side. The recognition score, disordered structure, and accuracy measurements are used to determine the assessment. Between these bounds, a market was drawn. DP pictures address a different kind of indicative assessment, emphasizing subjective inquiry and the separation of flaws. A paradigm that does not require alignment into aggregated units and allows for subjective segmentation via factual inquiry. The photos gathered would be very intriguing, particularly in location diagnostics or survey screenings. As a result, future research will concentrate on CNN engineering for multi-source PD identification and hyper parameter conversion for demonstration applications. This test bearing is a clear model at the moment inside the context of the future autonomous PD master. A writing audit is critical in every exploration project since it clearly indicates the necessity for work and foundation progression. It creates linked requests for upgrades from previous tests and permits unusual difficulties to occur, clearly characterizing all of the constraints in terms of the exploratory project's progress. This chapter examines a number of review papers on the detection and management of transmission line faults.

KEYWORDS

- **convolutional neural network**
- **covered conductor**
- **deep learning**
- **diagnostics**
- **high voltage insulation**
- **partial discharges**

REFERENCES

1. Sahoo, B. K., Sambit, P., Basanta, K. P., Baladev, B., Nimai, C. P., & Saloni, D., (2020). An artificial neural network for fault detection in an electrical power transmission system. *1–4 in the 2020 International Conference on Computational Intelligence for Smart Power Systems and Sustainable Energy (CISPSSE)*, IEEE.
2. Bouthiba, T., (2004). Fault location in EHV transmission lines using artificial neural networks. *International Journal of Applied Mathematics and Computer Science, 14*(1), 69–78.
3. Venkata, R. Y., Ravindhranath, B., Vijay, K. M., Surya, K. M., & Reddy, B. R., (2005). Detection and localization of transmission line faults using wavelet transformations (Coif let & Mexican hat). *Journal of Theoretical and Applied Information Technology.*
4. Karthikeyan, K., (2007). *Power System Failure Detection and Classification Using Wavelet Transformations and Adaptive Resonance Theory Neural Networks.*
5. Shaaban, S. A., & Takashi H., (2010). Transmission line faults classification using wavelet transform. *IEEE, 14th International Middle East Power Systems Conference (MEPCON'10).* Cairo University, Egypt.
6. Suhaas, B., & Ayyagari, (2011). *Artificial Neural Network-Based Fault Location for Transmission Lines.*
7. Tayeb, E. B. M., & Omer, A. A. A. R., (2011). Transmission line faults detection, categorization, and localization using artificial neural network. *IEEE International Conference and Utility Exhibition on Power and Energy Systems: Issues and Prospects for Asia (ICUE)* (pp. 1–5).
8. Abhijit, A. D., Kadu, A. N., & Rao, M. M., (2011). Intelligent control for locating fault in transmission lines. *International Journal of Instrumentation, Control & Automation (IJICA).* ISSN: 2231-1890.
9. Tayeb, E. B. M., (2013). Fault detection in power systems using artificial neural network. *American Journal of Engineering Research.*
10. Neeraj, P. S., Srivastava, R. K., & Vashishtha, P. K., (2014). Fault detection and isolation (FDI) via neural networks. *International Journal of Engineering Research and Applications* (Vol. 4, No. 1, pp. 81–86). ISSN: 2248-9622, (Version 1).
11. Smriti, K., & Dharmendra, K. S., (2014). Simulation of fault detection for transmission line protection using neural network. *International Journal of Science, Engineering and Technology Research (IJSETR), 3*(5).
12. Mohammad, A. D., (2014). Using network for types of fault in transmission lines. In: Adelian, R. S., (ed.), *International Journal of Engineering Research and General Science* (Vol. 2, No. 4).
13. Crowther, W. J., Edge, K. A., Burrows, C. R., Atkinson, R. M., & Woollons, D. J. (1998). Fault diagnosis of a hydraulic actuator circuit using neural networks—An output vector space classification approach. *Proceedings of the Institution of Mechanical Engineers, Part I: Journal of Systems and Control Engineering, 212*(1).

CHAPTER 18

EFFECT OF LOW AND HIGH CETANE FUEL BLENDS FOR DI DIESEL ENGINE PERFORMANCE CHARACTERISTICS

PEEYUSH DEWANGAN, SURAJ BANDHEKAR, RAM KRISHNA RATHORE, TEJASVI ANANT, SHAIK IQBAL ALI BASHA, and ASHOK KUMAR SHUKLA

Department of Mechanical Engineering, Rungta College of Engineering and Technology, Bhilai, Chhattisgarh, India

ABSTRACT

The energy crisis and environmental degradation are the main challenge for the scientist and researchers of the world due to growing population, industrialization and increased fossil fuel consumption. The proper disposal of solid waste tire is one of the very common and major problems in all over the world. One of the most efficient disposing ways is pyrolysis, which is environmental friendly and also yields useful products which may be used for the production of energy. Pyrolysis is the thermal degradation of solid wastes at high temperatures (300–900°C) in the absence of air to produce char, liquid oil or gas. The waste tire derived liquid (WTDL) can be a potential alternative fuel for CI engines but the main drawback is its lower cetane number, which is in the range of 25–30. In this context, a research work was carried out by the authors to study the effect of blending WTDL with a fuel, whose cetane number is greater. For this purpose WTDL was blended in different proportions with Jatropha biodiesel (JB) and diesel whose cetane numbers were higher. The WTDL and JME were blended in different percentage from 5% to 20% in a regular interval of 5% on a volume basis with mineral diesel. The aim of the current study is to investigate the

Computational Optimization, Modeling, and Simulation for Engineering Applications.
Anupam Shukla, Sourabh Rungta, Mohan Awasthy, and Rakesh Himte (Eds.)

combustion and performance characteristics of a DI diesel engine run on these blend and comparison with base fuel diesel. The behavior of the engine were evaluated in terms of ignition delay, maximum heat release rate, brake thermal efficiency, exhaust gas temperature. The test result revealed that, the behaviour of the engine run on the D80WTDL10JB10 (i.e. 80% diesel, 10% JB and 10% WTDL) blend was better than other blends considered in the study.

18.1 INTRODUCTION

The whole world is facing an energy crisis which affects directly or indirectly the sustainable development, such as economic and social, of any country. Energy is the basic requirement and major input for the economic development of any country. The fast depletion of fossil fuels has received attention from researchers and scientists throughout the world. Many attempts are being carried out worldwide to find alternatives to fossil fuels in order to reduce dependency on them. This chapter discusses energy sustainability, the world energy scenario, and the importance of alternative fuels for diesel engines. It also describes the different alternative fuels for diesel engines, their utilization, and their environmental impacts.

Alternative fuels are considered the fuels of the present and the future. Nowadays, many vehicles are switching over to alternative fuels worldwide. The worldwide fossil fuel crisis in the 1970s made many countries aware of its rapid depletion and environmental degradation [1]. The significant attention was focused by the scientists and researchers on the development of alternative fuel resources. The number of automotive has also amplified tremendously over the last decade, which has further increased the demand for environmentally friendly fuels. The potential alternative fuels for diesel engines are vegetable oils, biodiesel, ethanol, and some non-conventional fuels such as fuel derived through waste automobile tires, used transformer oil, waste wood, waste plastic, etc.

18.2 EXPERIMENTAL SETUP

18.2.1 GENERAL

This study was performed to find out whether a diesel engine can be run on non-conventional fuels. In this way, a single cylinder, four-stroke, direct injection (DI) diesel engine with a power rate of 4.4 kW was tested on various diesel-Jatropha biodiesel (JB)-waste tire-derived fluid (WTDL) combinations, and the results were compared to diesel and introduced in the same way. To achieve the fuel mixtures for the examination, the JB and WTDL were mixed in small amounts (5–20% by volume) in diesel by 90–60% by volume. To establish a baseline, the observation was taken first by running it on JB and then diesel, and subsequently with additional mixtures.

18.2.2 DESCRIPTION OF THE TEST ENGINE

A 4-stroke diesel engine is used to perform the experiment. It has a power output of 4.4 kW. The technical description of the test engine is taken from the previous literature [3]. Figures 18.1 and 18.2, respectively, show a snapshot of the test arrangement and a schematic graph. A thermocouple is attached to the engine for measuring the exhaust gas temperature (EGT). A dynamometer of the eddy current type is mounted to the engine to impart load. A burette is used to calculate the rate of fuel flow. The time difference between when the fuel passes through the general sensor and when it approaches the base sensor is recorded to estimate fuel usage. Using an air box, an orifice meter, and a U-tube manometer, air consumption was calculated.

FIGURE 18.1 Photograph of the experimental set up.

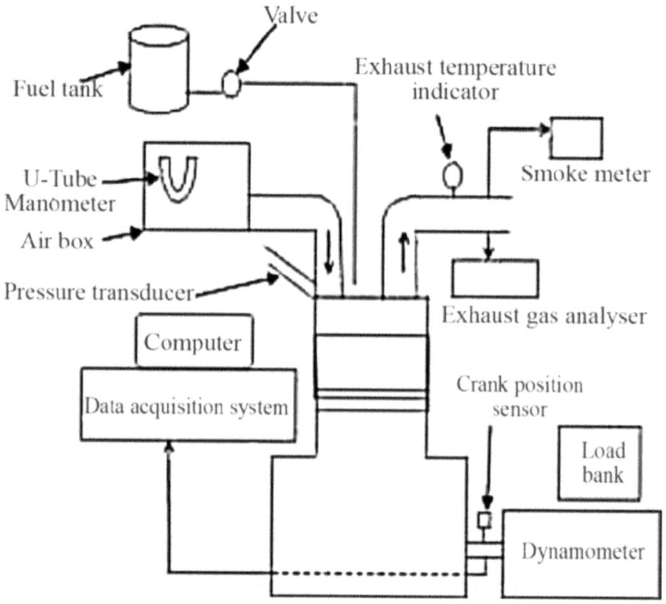

FIGURE 18.2 Schematic representation of experimental setup.

18.3 RESULTS AND DISCUSSION

18.3.1 GENERAL

The goal of this investigation was to see what would happen if a diesel engine were run entirely on non-petroleum fuel blends. Tests were performed in a naturally aspirated, single chamber, four-stroke, air-cooled, DI diesel engine with a rated power of 4.4 kW at a constant speed of 1,500 rpm on various diesel-JB waste tire-derived liquid (WTDL) mixes, with the results compared to diesel and presented in a similar manner. The WTDL and JB at low rates (5–20% at regular intervals of 5% by volume) were mixed with the leftover level of diesel to set up the fuel mixes for the examination. At first, diesel, JB, and the mixes were used in the engine without modifying its initial conditions as depicted in Table 18.1. The result of the analysis was compared with those of diesel-operated engines. The results obtained from the experiment with different mixes at different loads were used to improve the best mixes. The designations of the fuel blends and their compositions used in this study are listed in Table 18.1.

TABLE 18.1 The Designations of the Fuel Blends and Their Compositions

Fuel	JB (by Volume)	WTDL (by Volume)	Diesel (by Volume)
Diesel	–	–	1
JB	1	–	–
D90WTDL5JB5	0.05	0.05	0.90
D80WTDL10JB10	0.10	0.10	0.80
D70WTDL15JB15	0.15	0.15	0.70
D60WTDL20JB20	0.20	0.20	0.60

18.3.2 COMBUSTION PARAMETERS

18.3.2.1 IGNITION LAG (IL)

The IL for all the test fuels is shown in Figure 18.3. With the rise in the loading condition in the engine, the cylinder gas temperature also increases, leading to a decrease in IL [3]. The IL of JB, D90WTDL5JB5, and D80WTDL10JB10 blends is low with respect to diesel fuel. The reason for the low ignition lag (IL) is due to the high cetane number of JB [4].

The values of IL for diesel, JB, D90WTDL5JB5, D80WTDL10JB10, D70WTDL15JB15, and D60WTDL20JB20 blends are 11.5, 10.2, 11, 11.4, 11.9, and 12.7°C, respectively at full load. DP of diesel engine depends on the degree of fuel atomization and fuel viscosity [5]. With the increment in fuel viscosity the atomization will be reduced, and large size droplets will be generated. The more time will be consumed due to the large size of fuel droplets for the initiation of combustion, and which increases the delay period. This is the reason for longer IL for D70WTDL15JB15 and D60WTDL20JB20 blends as their viscosities are much higher compared to base fuel diesel.

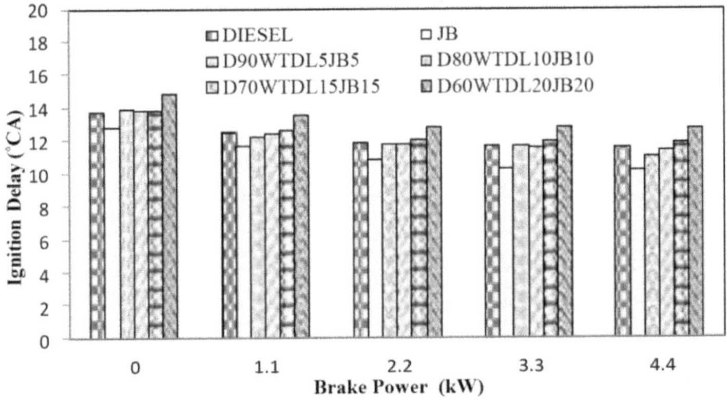

FIGURE 18.3 IL vs. BP.

18.3.2.2 HRR (HEAT RELEASE RATE)

Figure 18.4 depicts the relationship of HRR and crank angle for various test fuels under different loads. Comparable findings were made with different loads as well. Diesel has a longer DP, which causes the intensity of the burning phase mixture to increase, resulting in a greater HRR than the other fuels studied in this study. JB has a lower DP; hence it produces less heat. The cetane number of biodiesels was increased after the transesterification process [6].

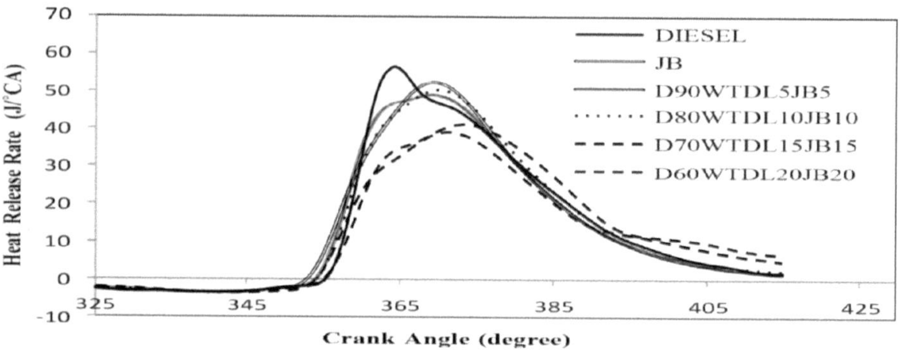

FIGURE 18.4 HRR vs. CA.

18.3.3 PERFORMANCE PARAMETERS

18.3.3.1 BTE (BRAKE THERMAL EFFICIENCY)

Figure 18.5 depicts correlation among BTE with BP for diesel, JB, and four fuel blends. The BTE is a significant term in engine performance measurement, and it refers to the ratio of power output to injected fuel energy. The product of the mass of the fuel injected per unit and its lower heating value yields the energy of the injected fuel [7]. The BTE is determined by a number of characteristics, including the fuel's calorific value, density, and viscosity, among others [8]. It was observed that the maximum BTE of 30.8% was produced by the engine during the entire engine operation.

The BTE of JB, D90WTDL5JB5, D80WTDL10JB10, D70WTDL15JB15, and D60WTDL20JB20 blends were noticed by about 28.9%, 29.5%, 29.9%, 28.6%, and 28.4% respectively at full load. From the above result, it can be noticed that when the volume of JB and WTDL is increased more than 10%, BTE of the blends starts decreasing. The blends with a higher percentage of JB and WTDL have lower BTE. The main reason behind this is the formation of spray is coarse and due to high viscosity of JB and WTDL than diesel improper atomization and poor mixture formation of blends takes place. The highest brake thermal efficiency was observed for D80WTDL10JB10, i.e., 29.9% at full load.

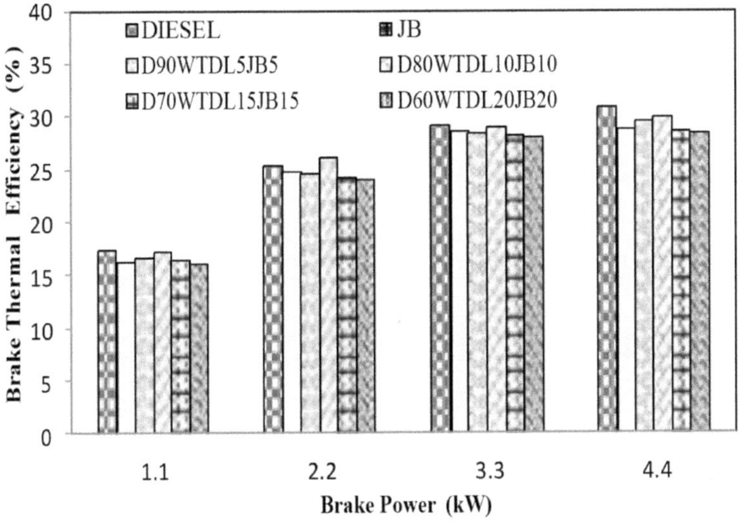

FIGURE 18.5 BTE vs. BP.

18.3.3.2 EXHAUST GAS TEMPERATURE (EGT)

Figure 18.6 shows the variation of EGT with respect to brake power for the engine using diesel, JB, D90WTDL5JB5, D80WTDL10JB10, D70WTDL15JB15, and D60WTDL20JB20 blends as test fuels. The EGT gives an idea about the amount of heat not utilized for producing power and going waste with the exhaust gases [9].

The values of EGT were 303, 320, 307, 315, and 322°C for diesel, JB, D90WTDL5JB5, D80WTDL10JB10, D70WTDL15JB15, and D60WTDL20JB20 blends, respectively. Figure 18.6 also indicates when the amount of JB and WTDL increases beyond 10% in blend, the EGT increased. At full load, the maximum value of EGT was for the D60WTDL20JB20 blend which was higher by about 19°C than diesel. This may be due to late burning of constituents of WTDL available in blend because WTDL contains some high boiling point compounds [10].

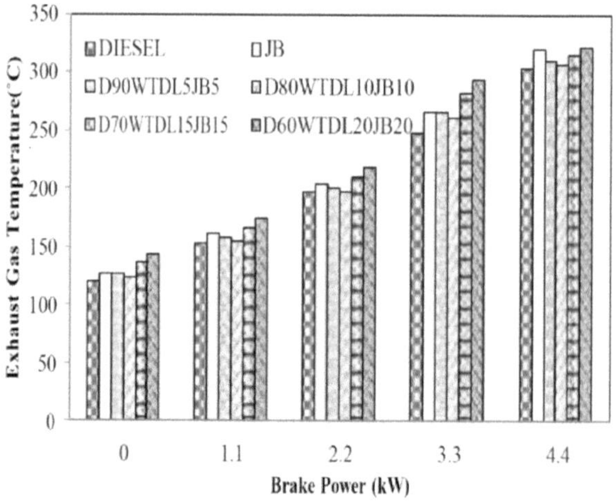

FIGURE 18.6 Exhaust gas temperature vs. BP.

18.4 CONCLUSION

By using the diesel-JB-WTDL blend fuels, the performance characteristics of the 4-stroke diesel engine are analyzed. The comparative analysis of outcomes for diesel and JB is represented for the engine characteristics. The following conclusions are obtained from the experiment:

- D80WTDL10JB10 and diesel fuel shows almost equal ignition lag at full load.
- The brake thermal efficiency for D80WTDL10JB10 is 29.9% and diesel is 30.8% at full load.
- EGT value for D80WTDL10JB10 is slightly higher in comparison with diesel at full load.

KEYWORDS

- **blending**
- **brake thermal efficiency**
- **diesel engine**
- **exhaust gas temperature**
- **heat release rate**
- **ignition lag**
- **jatropha biodiesel**

REFERENCES

1. *CIA World Fact Book,* (2014). https://www.cia.gov/library/publications/the-worldfactbook/geos/af.html (accessed on 21 December 2023).
2. Huang, J., Wang, Y., Qin, J. B., & Roskilly, A. P., (2010). Comparative study of performance and emissions of a diesel engine using Chinese Pistache and jatropha biodiesel. *Fuel Processing Technology, 91*(11), 1761–1767.
3. Flowers, W. L., Manson, R. K., & Kruger, C. H., (1975). In: *15th Symposium on Combustion* (p. 823). The Combustion Institute.
4. Li, J., Ou Chae, J., Park, S. B., Paik, H. J., Park, J. K., Jeong, Y. S., et al., (1997). *Effect of Intake Composition on Combustion and Emission Characteristics of DI Diesel Engine at High Intake Pressure*. SAE 970322.
5. Kong, S. C., Han, Z., & Reitz, R. D., (1995). *The Development and Application of a Diesel Ignition and Combustion Model for Multidimensional Engine Simulation.* SAE paper no. 950278.
6. Lai, J. Y. W., Lin, K. C., & Violi, A., (2011). Biodiesel combustion: Advances in chemical kinetic modeling. *Progress in Energy and Combustion Science, 37*, 1–14.
7. Suryanarayanan, S., Janakiraman, V. M., Rao, G. L., & Sampath, S., (2008). *Comparative Study of the Performance and Emission Characteristics of Biodiesels from Different Vegetable Oils with Diesel.* SAE Technical Paper.
8. Agarwal, A. K., Dhar, A., Gupta, J. G., Kim, W. I., Choi, K., Lee, C. S., & Park, S., (2015). Effect of fuel injection pressure and injection timing of Karanja biodiesel blends on fuel spray, engine performance, emissions, and combustion characteristics. *Energy Conversion and Management, 91*, 302–314.
9. Bhusnoor, S. S., Babu, M. G., & Subrahmanyam, J. P., (2007). *Studies on Performance and Exhaust Emissions of a CI Engine Operating on Diesel and Diesel Biodiesel Blends at Different Injection Pressures and Injection Timings.* SAE Technical Paper.
10. Jindal, S., (2011). Experimental investigation of the effect of compression ratio and injection pressure in a direct injection diesel engine running on Karanj methyl ester. *International Journal of Sustainable Energy, 30*(suppl. 1), S91–105.

CHAPTER 19

DETECTION OF MALICIOUS URLS USING MACHINE LEARNING

SHRUTIKA A. DESHMUKH,[1] SACHIN M. DANDAGE,[2] and P. M. JAWANDHIYA[3]

[1] *Research Scholar, 2nd Year M.E. in Department of Computer Science and Engineering, Pankaj Laddhad Institute of Technology and Management Studies, Buldhana, Maharashtra, India*

[2] *Assistant Professor, Department of Computer Science and Engineering, Pankaj Laddhad Institute of Technology and Management Studies, Buldhana, Maharashtra, India*

[3] *Principal, Pankaj Laddhad Institute of Technology and Management Studies, Buldhana, Maharashtra, India*

ABSTRACT

Malicious URL, a.k.a. malicious website, is a common and serious threat to cyber security. Malicious URLs host unsolicited content (spam, phishing, drive-by downloads, etc.) and lure unsuspecting users to become victims of scams (monetary loss, theft of private information, and malware installation), and cause losses of billions of dollars every year. It is imperative to detect and act on such threats in a timely manner. Traditionally, this detection is done mostly through the usage of blacklists. However, blacklists cannot be exhaustive, and lack the ability to detect newly generated malicious URLs. To improve the generality of malicious URL detectors, machine learning techniques have been explored with increasing attention in recent years. This article aims to provide a comprehensive survey and a structural

Computational Optimization, Modeling, and Simulation for Engineering Applications.
Anupam Shukla, Sourabh Rungta, Mohan Awasthy, and Rakesh Himte (Eds.)

understanding of Malicious URL Detection techniques using machine learning. We present the formal formulation of Malicious URL Detection as a machine learning task, and categorize and review the contributions of literature studies that address different dimensions of this problem.

19.1 INTRODUCTION

The World Wide Web has become an important base for using the web. Some advanced technology has become blind to a few enhanced strategies to attack on rip-off customers. In this, there are some attacks on websites regarding duplicate goods, financial online fraud with the aid of tricking users into relevant use records, user credentials, user important passwords, and many more. So, these activities need to stop and save the users from irrelevant responses. There are various types of attacks that can be done through the URLs, like SQL injections, as well as by using the phishing technique. It is possible to steal user information in this tree. Various cyber security branches are working and keeping an eye on the execution of such activities. But, as per the traditional approaches seen, there is only the execution done on traditional approaches like checking the black links of the URL. In this condition, we can evaluate the content of URLs so that the content can be evaluated and operated. So, the proposed mechanism can maintain the complete process of the execution in ethical content-based evaluation can work out [1–19].

19.2 EXPERIMENTAL METHODS AND INFORMATION

19.2.1 MALICIOUS URL

Malicious URL is a URL that is created for malicious purposes, intending to download any kind of malicious program, which can include it in spam or steal some user data or improve its condition on search engines using Blackhat SEO techniques.

Among the many cyber threats, cybercrime plays an important role in cyber-attacks and scams. Malicious URLs can be sent to users by mailing, messages, browser pop-ups or ads. The result can be downloaded by viruses such as malware, spyware, ransomware, accounts at risk, as well as headaches that threats include. It should be proof that being aware of what a Malicious URL is, and how it can secure email.

Simply put, a URL recipient can be like a criminal who invites you into a system. Their invitation may promise you good food or other things, and you may visit, but you do not know what will actually happen until you get inside. Who knows in the best-case scenario, there may be a more aggressive invitation that could mean that your wallet information will be stolen. It can also lead to information capture. We see that Malicious URLs are a major part of the many online security threats. They are aiding hackers use:

- Phishing is one of the ways by which user information can be stole.
- Mouse trick is a very easy technique to start the installation of malware into the machine, which may affect the system.
- Most of the phishing techniques are used for the manipulation and tracing of the users' records.

19.2.2 DETECT MALICIOUS URL BY MACHINE LEARNING METHOD

Methods for Learning Machine use multiple URLs as training data and based on statistical structures read the guesswork function to classify a URL as malicious or dangerous. This gives the ability to create new URLs as opposed to blocking methods. The main requirement for training a machine learning model is the availability of training data. In the context of the acquisition of a malicious URL, this will go hand in hand with a set of URL numbers. General electronic learning can be classified as supervised, unsupervised, and semi-supervised, with the availability of training data labels, the absence of labels, and the labeling of a limited portion of training data, respectively. Labels are associated with information that the URL is malicious or harmless.

These methods attempt to analyze the URL information and its associated websites or webpage, by extracting a good URL attribute representation, and training a predictive model in your training data for both malicious and malicious URLs. There are two types of attributes that can be used – vertical attributes and flexible attributes. In static analysis, we perform web-based analysis based on available information without using a URL. Extracted features include lexical features ranging from a unit of URLs, host information, and sometimes even HTML and JavaScript content. Since no execution is required, these methods are safer than others.

These approaches try to analyze the information of a URL and its corresponding websites or web page, by extracting good attribute representation of URLs, and training a prediction model on training data of both malicious and benign URLs. There are two types of functions available: static functions and dynamic functions. Static analysis does not run URLs, but rather performs web page analysis based on available information.

19.3 SYSTEM ANALYSIS AND DESIGN

19.3.1 PROBLEM ANALYSIS

This chapter addresses the concept or issues related to the previously existing systems and their limitations over their implementations. It also focuses on how the proposed system will overcome the drawbacks. From what is stated in the introduction, Blocked listing methods have some limitations, and it seems easy to go through them, especially since the block list does not help predict new URLs. Every time a new URL is introduced, it needs to be looked up in the existing database of malicious URLs. The new URL will be checked for all the previously known malicious URLs in the blacklist. Whenever a new malicious URL is detected on the system, the update should be blacklisted. This method is iterative, time-consuming, and computationally intensive with an ever-increasing number of new URLs.

19.3.2 SOFTWARE REQUIREMENTS

- Python IDLE 3.0;
- MySQL.

19.3.3 LANGUAGE USE

- **Front End:** Tkinter.
- **Back End:** MySQL.
- **Languages:** Python, SQL.

19.4 RESULT ANALYSIS ON SOFTWARE SYSTEM (FIGURE 19.1)

FIGURE 19.1 Flow diagram malicious URL determination.

19.4.1 INDEX

Index consists of a login window where the user has to log in if he/she is already registered or if they are a new user then they have to register themselves (Figure 19.2).

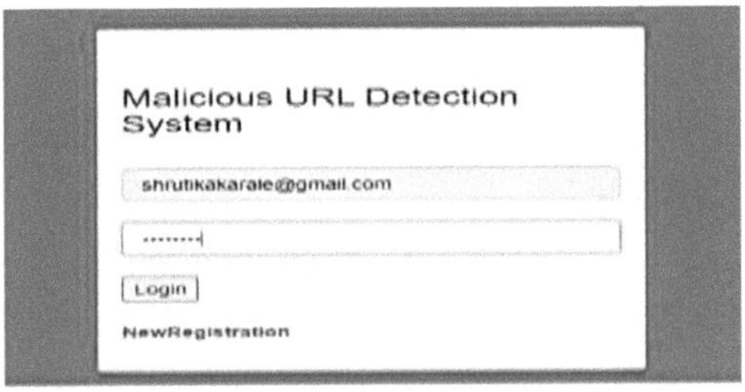

FIGURE 19.2 Login of malicious URL detection system.

19.4.2 USER DASHBOARD

In the User Dashboard page, where we see certain options, such as search for add stop words, add categorical words, view searches, view results, blacklisted links, logout (Figure 19.3).

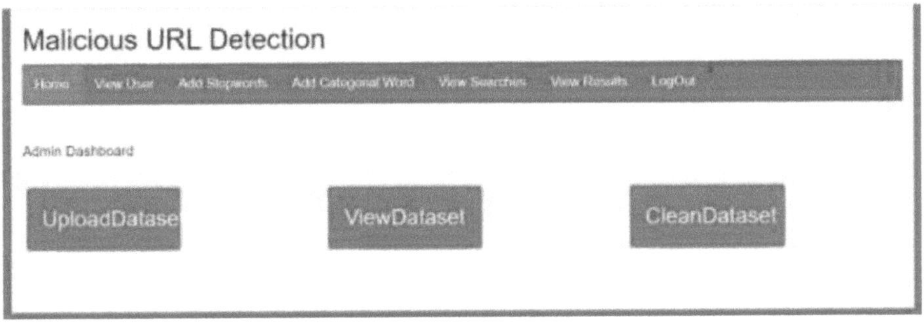

FIGURE 19.3 Dashboard of malicious URL detection system.

19.4.3 SEARCH FOR NEW LINKS

In the search for a new link option, we have to enter a new link for the detection of malicious URLs (Figure 19.4).

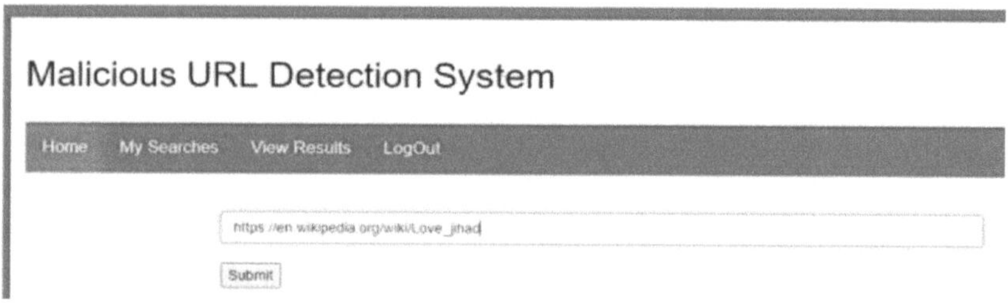

FIGURE 19.4 Enter URL to search for malicious URL detection system.

19.4.4 CRAWLED DATA

As soon as we enter the link and click on the search button, the data is extracted from the link using the web crawler technique. Web Crawler is a technique used to retrieve data from any website. Web crawlers are often used by search engines for web indexing (web spidering) (Figure 19.5).

FIGURE 19.5 Crawled data from URL.

19.4.5 MALICIOUS URL DETECT AND RESULT

The next step is the extraction of data from URL and applying data analytics. After applying the data analytics, the Term Frequency Determination is done which shows how many malicious words are detected. To generate a result for malicious URL, we have to click on the generate result button. The result is generated as shown in Figure 19.6.

FIGURE 19.6 Malicious detected of URL.

19.4.6 BLACKLISTED LINKS

When we pass a link to detect a malicious URL, the data is sent and then it is compared with the keywords. If the keywords match the given ratio, i.e., 50%, then that link is blacklisted, which is shown in the blacklisted link option (Figure 19.7).

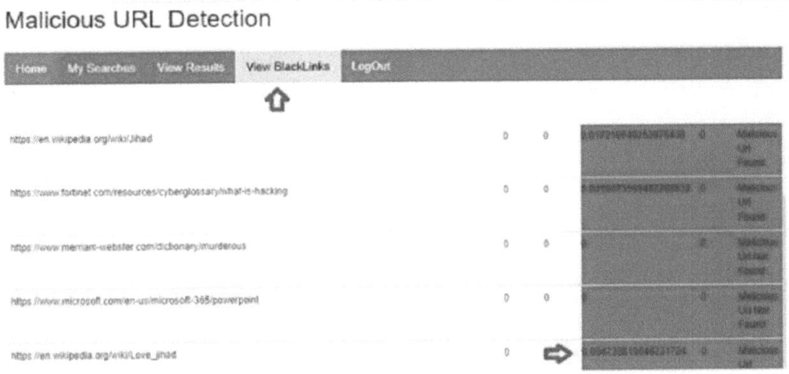

FIGURE 19.7 Malicious URL found on BlackLinks.

19.4.7 LOGOUT

Logout button is used to log you out from malicious URL detection system.

19.5 CONCLUSION

In the proposed web crawling-based web page mining, the performance is effective in finding malicious content compared to others. This makes the proposed methodology highly effective compared to previous methods. After experimenting on the URL reputation dataset, we were able to examine a distributed machine learning model using Logistic Regression and SVM algorithms. We concluded that when working with extremely small data, traditional machine learning algorithms were more efficient and computed much faster. This is because Spark and Java introduce overhead and some complexity is lost for small computations. However, as we gradually increase the amount of data we are working with, Spark outperforms traditional machine learning methods. During the experiment, there was a loss of node availability, causing Spark to work offline. The server and worker were connected to the same node.

KEYWORDS

- **blacklisted links**
- **crawled data**
- **ensemble learners malicious URLs**
- **index**
- **machine language**
- **machine learning**
- **software requirements**

REFERENCES

1. Sahoo, D., Liu, C., & Hoi, S. C. H., (2017). *Malicious URL Detection Using Machine Learning: A Survey.*CoRR, abs/1701.07179.
2. Khonji, M., Iraqi, Y., & Jones, A., (2013). Phishing detection: A literature survey. *IEEE Communications Surveys & Tutorials, 15*(4), 2091–2121.
3. Cova, M., Kruegel, C., & Vigna, G., (2010). Detection and analysis of drive by-download attacks and malicious JavaScript code. In: *Proceedings of the 19th International Conference on World Wide Web* (pp. 281–290). ACM.
4. *Internet Security Threat Report (ISTR) 2019–Symantec.* https://www.symantec.com/content/dam/symantec/docs/reports/istr-242019-en.pdf (accessed on 21 December 2023).
5. Sinha, S., Bailey, M., & Jahanian, F., (2008). Shades of grey: On the effectiveness of reputation-based—Blacklists. In: *Malicious and Unwanted Software; Malware 2008 – 3rd International Conference* (pp. 57–64). IEEE.
6. Ma, J., Saul, L. K., Savage, S., & Voelker, G. M., (2009). Identifying suspicious URLs: An application of large-scale online learning. In: *Proceedings of the 26th Annual International Conference on Machine Learning* (pp. 681–688). ACM.
7. Leo, B., (2001). Random forests. *Machine Learning, 45*(1), 5–32.
8. Thomas, G. D., (2000). Ensemble methods in machine learning. *International Workshop on Multiple Classifier Systems* (pp. 1–15). Cagliari, Italy.
9. *Developer Information*, https://www.phishtank.com/developer_info.php (accessed on 21 December 2023).
10. https://urlhaus.abuse.ch/downloads/csv/ (accessed on 21 December 2023).
11. https://www.kaggle.com/antonyj453/urldataset#data.csv (accessed on 21 December 2023).
12. Immadisetti, N. V. D. N., Manamohana, K., & Rohit, V., (2019). Detection of malicious URLs using machine learning techniques. In: *Proceedings of the International Journal of Innovative Technology and Exploring Engineering (IJITEE)* (Vol. 8, No. 4S2). ISSN: 2278-3075.

13. Doyen, S., Chenghao, L., & Steven, C. H. H., (2019). *Malicious URL Detection Using Machine Learning: A Survey, 1*(1), 37.
14. Dataset URL. http://downloads.majestic.com/majestic_illion.v(accessed on 21 December 2023).
15. https://drive.google.com/file/d/13G_Ndr4hMFx_qWyTEjHuOyJmHFWD0Gud/view?fbclid=IwAR0SLVCrvjHHGmoHZH97nXN3BmDMY7jG4SOsKZYLAZjTFgeoJADfli64-g (accessed on 21 December 2023).
16. Dharmaraj, R. P., & Jayantrao, B. P., (2015). Survey on malicious web pages detection techniques. In: *Proceedings of the International Journal of u- and e- Service, Science and Technology* (Vol. 8, No. 5, pp. 195–206). http://dx.doi.org/10.14257/ijunesst.2015.8.5.18.
17. Dharmaraj, R. P., & Jayantrao, B. P., (2018). Feature-based malicious URL and attack type detection using multi-class classification. In: *Proceedings of The ISC Int'l Journal of Information Security* (Vol. 10, No. 2, pp. 141–162).
18. Heartfield, R., & Loukas, G., (2015). A taxonomy of attacks and a survey of defense mechanisms for semantic social engineering attacks. *ACM Computing Surveys (CSUR), 48*(3), 37.
19. Tao, Y., (2014). *Suspicious URL and Device Detection by Log Mining*. Ph.D. dissertation. Applied Sciences: School of Computing Science.

CHAPTER 20

A REVIEW OF THERAPEUTIC APPROACHES AND PATHOGENESIS OF DIABETIC RETINOPATHY

SHREYA SINGH and ROSHNI SHARMA

Faculty of Pharmacy, Rungta College of Pharmaceutical Science and Research, Raipur, Chhattisgarh, India

ABSTRACT

This review chapter summarizes the types, pathological features, and treatment of diabetic retinopathy (DR). It can be clearly seen that there is an enlargement in the radius of use of non-insulin anti-hyperglycemic agents in the treatment of type 2 diabetes. Globally, the number of type 2 diabetes cases has increased, and the complications associated with DR have also increased. It is of great significance for ophthalmologists to be updated with the newly introduced agents that have an effect on DR and diabetic macular edema. DR is a cardiac complication in diabetic patients; it has gone undetected and unnoticed so far until irreversible, serious damage and even blindness have occurred in the patient. It has been proven that effective detection and screening of this disease reduces the risk of blindness. With a special emphasis on the initial stages of diabetes, increased awareness of the complex neuronal, glial, and cardiac abnormalities that progressively and constantly affect retinal function. Few drugs like Thiazolidinediones (The only drug available currently is pioglitazone) have been linked with incidents of worsening condition in diabetic macular edema; however, it is believed that the range so far is low. In general, glucagon-like peptide-1 (GLP-1) agonists are much more beneficial, except semaglutide as it is associated

Computational Optimization, Modeling, and Simulation for Engineering Applications.
Anupam Shukla, Sourabh Rungta, Mohan Awasthy, and Rakesh Himte (Eds.)

with an inclining rate of diabetic retinopathy complications (DRC). DPP4 inhibitors and Biguanides are also used for its treatment. Here, it is a presentation of a review of the current understanding and scenario of new insights into pathogenesis in DR, as well as clinical treatments for DR patients.

20.1 INTRODUCTION

There are mainly three layers of nerve cell bodies diverged by two plexiform (region in which the neurons make their interconnections) layers of synapses in the human retina. Rod and cone photoreceptor cell bodies are located in the outer nuclear layer. Bipolar, horizontal, and amacrine cells (unipolar nerve cells) are present in the inner nuclear layer, and ganglion cell bodies and displaced amacrine cells (unipolar nerve cells) from the ganglion cell layer. Photoreceptor cells connect to second-order neurons, primarily bipolar and horizontal cells, and second-order neurons get connected to ganglion cells in the inner plexiform layer (region in which the neurons make their interconnections). The intraretinal vasculature is an end artery, multilayered capillary network perfusing the inner retina [1]. Diabetic retinopathy (DR) is a carotid complication of diabetes which is sight-threatening and severely affects the retina. It is the most common complication seen in diabetes and a main or leading cause of sight loss in working-aged adults. All persons having diabetes are at great risk of developing retinopathy disease, it has been seen that persons living with type 1 diabetes (T1DM) are at a higher risk of getting this disease as compared to persons suffering from type 2 diabetes (T2DM).

20.1.1 TYPES

Founded on the appearance or non-appearance of blood vessels on the retina that are abnormal, it has been classified into two categories:

- Non-proliferative diabetic retinopathy (NPDR); and
- Proliferative diabetic retinopathy (PDR) [2].

20.1.1.1 NON-PROLIFERATIVE DIABETIC RETINOPATHY (NPDR)

The formation of microaneurysms (saccular enlargement of the venous end of a retinal capillary), i.e., outwardly bloating of the capillary wall, is seen in the early morphology of nonproliferative diabetic retinopathy (NPDR). It is located on the temporal side of the fovea and is usually asymptomatic when it first arises; they may cause intraretinal hemorrhages if ruptured. It can be detected by an ophthalmoscope. Fluorescent angiography is helpful in revealing leakage, which confirms the cause of macular edema. It has been considered a classical symptom of ischemia and a predictor of imminent progression towards proliferative retinopathy (PR).

Nerve fiber layer has microinfarcts, known as cotton wool spots. Sometimes it is an indication of non-controllable arterial hypertension.

20.1.1.2 PROLIFERATIVE DIABETIC RETINOPATHY (PDR)

In the retinal capillary bed, blood flow becomes impaired and spreads over the retinal area, confirming the development of proliferative diabetic retinopathy (PDR). Neovascularization initiates at the papilla due to ischemia (neovascularization of the disc (NVD)). Retinal proliferation in blood vessels can be seen as a futile and vain attempt to compensate for ischemia, as it forms new vessels initially at the papilla, then on the retina, and finally on the iris. Neovascularization in the papilla and retina can lead to hemorrhages in the epiretinal and subhyaloid vitreous areas and may form membranes and cords on the surface of the retina. Eventually, contraction of these structures can cause detachment of the retina or fractional macular edema, both of which can result in blindness. Neovascular glaucoma is the most severe complication of DR. It involves the growth of newly formed vessels from the pupil into the chamber angle, which obstructs the outflow of aqueous humor.

20.2 PATHOGENESIS

20.2.1 EARLY MICROVASCULAR LESIONS

These vascular lesions, which are spotted with the initiation of retinopathy, include the genesis of aneurysms in capillaries, vanishing of pericytes from capillaries possessing endothelial cells, non-perfusion and eradication in capillaries and small arterioles, gradually thickening in the vascular basement membrane, and associated changes like vessel leakage, exudates, and hemorrhage. In normal eyes of a person not suffering from diabetes, it is difficult to establish the sequence in which the initial lesions appear precisely. A patchy distribution in the retina can be seen due to abnormality, an occasional aneurysm confounded because of early changes, pericytic ghost, or obliterated capillary that may be found apparently. Pericyte loss initially becomes an immoderate regularity of degeneration of pericytes and pericytic ghost. In diabetes, endothelial cells also degenerate, early changes in the endothelial cell population are more difficult to define as enough autotrace is left behind [3].

20.2.1.1 STAGE 1: MILD NPDR

Patients with mild NPDR must have an examination (dilated eye) every 12 months. Within the span of one year, mild NPDR will start converting to PDR (5% risk). For easier future comparison, Fundus photography can be used. Patients having mild NPDR need no referral to a specialist unless they are concerned about or confirmed a diagnosis. Discussing the findings with patients is important, especially with those who have been recently diagnosed with diabetes, to make sure that they understand that this may indicate sometime early end organ damage. Their blood sugar and diet should be monitored to avoid further complications.

20.2.1.2 STAGE 2: MODERATE NPDR

Patients experience hemorrhage in retinal quadrants (one to three). Patients with moderate NPDR should visit for an eye examination after every 6 to 8 months. There is a (12%–27%) risk of developing PDR within one year. For such cases, Fundus photography is suggested. These patients do not need a referral to an oculist unless diagnosed thoroughly. Educating the patients about the disease is important.

20.2.1.3 STAGE 3: SEVERE NPDR

The patients experience hemorrhages in the intra-retinal area, >20 in each quadrant, beading of venous in quadrants (two or more), or having an IRMA in one or more quadrants. If such findings are seen in the absence of neovascularization, it indicates PDR. Any initial neovascularization should be detected using both macular OCT and fluorescent angiography in patients suffering from severe NPDR. Referring to an ophthalmologist is recommended. Patients should be monitored with Fundus examination (dilated eye) every 3 to 4 months. Patients having severe NPDR may have a risk of 52% of developing PDR within a span of one year, so it becomes mandatory to discuss the importance of controlling blood sugar levels and observation with them. They are at great risk of progression of disease and permanent loss of vision due to diabetes and are likely to experience neuropathy elsewhere at this point.

20.2.1.4 STAGE 4: PROLIFERATIVE DIABETIC RETINOPATHY (PDR)

The patients with NPDR that has progression towards PDR exhibit pre-retinal hemorrhage or neovascularization of the disc. These patients should immediately need referral to an ophthalmologist for further treatment. Laser retinal photocoagulation is used to treat peripheral neovascularization. Anti-VEGF intravitreal injections may also often be received by them. Until their diabetes stabilizes, they should be monitored monthly by an ophthalmologist. Thereafter, they can be seen after every 6 to 12 months [4].

20.3 TREATMENT

Plainly, it is evident that the risk of macrovascular and microvascular complications can be reduced by initial control of glucose, and maintaining sustained and better metabolic control is the medicinal treatment to hold back the enlargement and to delay the progress in DR. The period of untimely glucose control is always contemplated as one of the optimal factors in the end result of the high control, and this also increases the value of starting appropriate treatment protocol for patients having diabetes, particularly in patients who have type 2 diabetes, during diagnosis, to retard further progress in DR and other complications. However, high control in glycemia is also related to good continuation cost and side effects like hypoglycemia, mainly in older patients, patients having acute cognitive impairment, patients having dementia, and also those who experience high incidences of coronary syndrome and mortality.

Modifications in epigenetics can be stabilized and passed on to the next generations, and these modifications ultimately regulate transcription. During the reversal period, direct inhibition of DNMT activation by its inhibitor helps in initiating activation of DNA methylation machinery (DNA mm) [5].

20.3.1 GLUCAGON-LIKE PEPTIDE 1 (GLP-1) AGONISTS

Experimental research has shown potential and protective effects of glucagon-like peptide-1 (GLP-1) agonists on the retina of patients with diabetes via the function of improved blood-retina barrier (IBRB) and decreased neuronal apoptosis. Administration of drugs of this class through the Intravitreal route suggests that it is protective to the retina by declining glutamate levels through upregulation of GLP-1 receptor and glutamate-aspartate transporter (GLAST).

In the early experimental phase of DR, a long-acting GLP-1 agonist shows neuroprotective effects by reducing the growth factor of the placenta and expressing intercellular adhesion molecule-1, which maintains the integrity of the IBRB. Administration of drugs such as liraglutide through the systemic route prevents neurodegeneration of the retina by reducing extracellular glutamate and increasing pro-survival pathways for signaling. Identical neuroprotective effects can be seen using lixisenatide and exenatide, independent of glucose reduction. This suggests that the activation of the GLP-1 receptor itself has neuroprotective effects on the retina in the case of DR.

20.3.2 DPP4 INHIBITORS

DPP4 is an enzyme which is behind the fast deterioration of GLP-1. However, as a consequence, DPP4 inhibitors have been developed which delay the breakdown of incretins, resulting in an extension of their action. The advantage of DPP4 inhibitors over GLP-1 agonists includes the administration of the drug orally rather than in parenteral form, which also avoids side effects such as nausea and vomiting associated with GLP-1 agonists. Experiments on DPP4 inhibitors indicate their protective effects on DR.

20.3.3 BIGUANIDES

Metformin has been demonstrated as a cardioprotective agent independent of glycemic control. It has shown anti-inflammatory and anti-angiogenic effects by increasing the level of thrombospondin-1, which decreases the concentration and activation of plasminogen activator inhibitor-1 and thus increases fibrinolytic activity. Since inflammation and angiogenesis are factors in the development of retinopathy, metformin may have protective effects against retinopathy. However, its effects on glycemic control need to be examined [6].

20.3.4 SURGERY

Subhyaloid hemorrhage, non-resorption, vitreous hemorrhage, fractional detachment of retina, ghost-cell glaucoma and fractional macular edema are indications of PPV (pars plana vitrectomy). The proper repositioning of retina, removing turbid vitreous body and optimal laser photocoagulation treatment is enabled by PPV. Operative time has been shortened by technical refinements and preclude the need for suturing in recent years. The diameter of the instruments has been reduced to 1.0 to 0.6 mm that is currently introduced through trocars. It is now possible to maintain at least basic vision with surgery in patients with advanced proliferative DR[7].

20.3.5 ANTIVEGF THERAPY

This treatment incompletely resolves edema (in 20%–40% of eyes). Nowadays ranibizumab and aflibercept are approved by the Food and Drug Administration (FDA) for treating DR, moving from initial to a higher VEGF binding affinity drug (sequence of VEGF binding affinity is as: Bevacizumab < ranibizumab < aflibercept) has been studied. An approach for treating the incomplete response is to shift an anti-VEGF to a corticosteroid or in a combination therapy. This improves the thickening in the maculae; however, the addition of VA gains is variable [8].

KEYWORDS

- **clinical trials**
- **diabetic retinopathy**
- **glucagon-like peptide 1**
- **microvascular lesions**
- **ophthalmologist**
- **pathophysiology**
- **proliferative diabetic retinopathy**

REFERENCES

1. Lechner, J., & O'Leary, O. E., (2017). The pathology associated with diabetic retinopathy. *Vision Research* (Vol. 4, No. 3, pp. 1–3). Elsevier.
2. Kashim, M. R., (2018). Diabetic retinopathy screening: A systematic review on patients' non-attendance. *International Journal of Environmental Research and Public Health, 15*, 157.

3. Engerman, L. R., (2017). *Perspectives in Diabetes Pathogenesis of Diabetic Retinopathy, 38*, 1203–1204.
4. Koetting, C., (2019). The four stages of diabetic retinopathy. *Cover Focus Diabetic Eye Disease (Modern Optometry).*
5. Kowluru, A. R., (2018). Diabetic retinopathy, metabolic memory, and epigenetic modifications. *Vision Research* (Vol. 2, No. 11, pp. 35–36). Elsevier.
6. Saw, M., (2019). New anti-hyperglycaemic agents for type 2 diabetes and their effects on diabetic retinopathy. *Nature* (pp. 1–6). Springer.
7. Kollias, N. A., (2018). Diabetic retinopathy early diagnosis and effective treatment. *Deutsches Ärzteblatt International (Medicine), 107*(5), 76–80.
8. Stewart, W. M., (2019). Treatment of diabetic retinopathy: Recent advances and unresolved challenges. *World Journal of Diabetes, 7*(16), 336–338.

CHAPTER 21

EXPLORING HEAT DYNAMICS: ANALYZING TURBINE BLADE PERFORMANCE

N. ARJUN,[1] MOON BANERJEE,[2] MOHAN AWASTHY,[3] KANTILAL RANE,[4] and M. SREEKANTH[5]

[1]*Associate Professor, Malla Reddy Institute of Engineering and Technology, Hyderabad, Telangana, India*

[2]*Associate Professor, Koneru Lakshmaiah Educational Foundation, Deemed to be University, Hyderabad, Telangana, India*

[3]*Principal and Professor, Rungta College of Engineering and Technology, Bhilai, Chhattisgarh, India*

[4]*Professor, Koneru Lakshmaiah Educational Foundation, Deemed to be University, Hyderabad, Telangana, India*

[5]*Assistant Professor, Malla Reddy Institute of Engineering and Technology, Hyderabad, Telangana, India*

ABSTRACT

Turbine blades are critical components in gas and steam turbines, responsible for extracting energy from the high-temperature, high-pressure working fluid. Due to their operational environment, these blades experience significant thermal loads, leading to complex heat transfer phenomena and potential thermal stress issues. This work presents a thermal analysis of turbine blades, focusing on modeling, and validation followed by optimization strategies. The analysis aims to understand the thermal behavior of the

Computational Optimization, Modeling, and Simulation for Engineering Applications.
Anupam Shukla, Sourabh Rungta, Mohan Awasthy, and Rakesh Himte (Eds.)

blade, predict temperature distributions, and identify critical locations for potential failures.

21.1 INTRODUCTION

Turbine blades of a turbine in a gas turbine engine are held onto a disk or hub which holds them. The fluid enters the compressor and gets compressed, resulting in an increase in temperature and pressure. Air is combusted in the combustion chamber, which is located between compressor stages and turbine stages, and in turn, the air temperature is further increased to a great amount. The turbine stages are then exposed to high temperature and high-pressure exhaust gases. The pressure and temperature of the air decrease as it passes out of the combustion chamber, and the air is directed onto the turbine stages where the energy is extracted and transferred to the compressor stages.

Engines having more turbine stages have a high bypass ratio. The design of turbine blades has a great effect on the number of turbine stages. Many gas turbine engines have a two-spool design, i.e., a high-pressure and low-pressure spool. A few engines also have three spools, introducing an intermediate pressure spool. The turbine exposed to hot air and high-pressure air is called the high-pressure turbine, and the one exposed to low temperature and pressure is called the low-pressure turbine. The fabrication of turbine blades with a pattern of high pressure and low pressure, along with a variety of material combinations, is led by the difference in said conditions, even though the thermodynamic and aerodynamic principles are the same.

Inside the gas turbine, the turbine blades are exposed to a dynamic atmosphere. High temperatures and peak damping variations are experienced by the turbine blades, which can further stress the blades of the turbine or the entire engine. Thus, an exponential number of parameters should be considered while designing turbine blades.

The old engines have one of the controlling features as the tailoring of the materials, which can further control the high temperature section of the engine. Thus, the requirement of the better tailored materials became the main criteria for the upcoming engines having designed blades. The availability of the new materials, their combination, like alloy, composite, nanomaterials provide much better options to control the uncertainty.

Advancement of super alloys and temperature capability of turbine blades was increased to a great extent by vacuum induction melting in the 1950s. Turbine blades' performance was increased by using improved alloys made by the hot isostatic pressing method.

Sandip et al. [1] performed the thermal hydraulic analysis of an axial gas turbine for the sake of service life tenure. The study has been performed with the application of high pressure and temperature on the turbine blade with the help of simulation software. Harjot [2] has reviewed the numerical analysis of the turbine blade. The idea of intense temperature and pressure is further explored on the turbine blade and the final outcome is validated with the already available data. Farhad et al. [3] have conducted numerical analysis on gas turbine made of ZrB_2 material. Simulation software has been used to access the thermal analyzes of turbine stator blade made of ZrB_2. Kourosh et al. [4] performed the analysis of turbine blade with different materials, so that TiB_2 can act as a new material. The stresses near the maximum temperature region have been evaluated, keeping different parameters like pressure and different boundary conditions. Sushila et al. [5] have performed the failure analysis of gas turbine blade tip cracking in the thermal power plant. It is a practical study performed in order to learn the failure of turbine blade tip, which gets deteriorated by the continuous usage. Irfan et al. [6] have performed the numerical analysis of nickel chromium and titanium carbide gas turbine blade cooling. In this research work, the cooling factor of turbine cooling blade has been focused on as a high priority for the smooth functioning of the turbine blade. Three different cooling methods have been identified and tested for the improvement of film cooling. Tao Yang et al. [7] investigated the crack propagation in the martensitic stainless steel turbine blade in the power plant.

A significant advancement for alloy is the creation of unit crystal production techniques. By reorienting the grain boundaries in a particular direction, these techniques significantly boost strength against fatigue and creep. Figure 21.1 shows the turbine blade which is being taken into consideration for the analysis purpose. The diameter and length of the blade are 1308.5 mm and 117 mm.

FIGURE 21.1 Turbine blade.

The number of turbine stages can have a great effect on how the turbine blades are designed for each stage. Many gas turbine engines are twin spool designs, meaning that there is a high-pressure spool and a low-pressure spool. Other gas turbines use three spools, adding an intermediate pressure spool between the high- and low-pressure spool. The high-pressure turbine is exposed to the hottest, highest-pressure air, and the low-pressure turbine is subjected to cooler, lower pressure air. That difference in conditions leads the design of high pressure and low-pressure turbine blades to be significantly different in material and cooling choices even though the aerodynamic and thermodynamic principles are the same. Turbine blades are subjected to very strenuous environments inside a gas turbine. They face high temperatures, high stresses, and a potentially high vibration environment. All three of these factors can lead to blade failures, which can destroy the engine, and turbine blades are carefully designed to resist those conditions. The high temperatures can also make the blades susceptible to corrosion failures. Finally, vibrations from the engine and the turbine blade can cause fatigue failures.

21.2 DESCRIPTION OF PROBLEM

Thermal stress analysis is performed on a turbine blade with pressure and temperature applied to it. Three different materials are considered, and thermo-structural analysis was performed. The materials considered are N155, Hastelloy X, and Inconel 625 alloy. The above materials are used in view of their physical, chemical, and structural activities, which result in inherent behavior with respect to strength, toughness, fatigue, and resistance towards corrosion. Mechanical properties of N155, Inconel 625, and Hastelloy X alloy. Figure 21.2 shows the different material properties that are being used in the research work below.

Properties	**Units**	**N155**	**Incone 16625**	**Haste alloy X**
E	Pa	143E9	15E09	144E09
p	Kg/cu m	8249	8400	8300
K	w/m-K	20.0	10	25
μ	—	0.344	0.331	0.348
α	E-06/°C	17.7	15	16
CP	J/KgK	435	410	450
Melting	°C	1354	1350	1380
Yield stress	MPa	550	1030	360

FIGURE 21.2 Material properties. [L: Length; D: diameter of shaft; α: coefficient of thermal expansion; μ: Poisson's ratio; σ: stress; E: Young's modulus; δ: deflection; ρ: density; and K: thermal conductivity].

21.3 FINITE ELEMENT METHOD

This is the latest analytical method to find the approximate solution for the problem, so that the problem can be analyzed properly before going for experimental solution. The FEM gives the ease to opt for the suitable element, followed by meshing and then after load application, the results can be scrutinized for further analysis.

It is more suitable for a complex geometry. Figure 21.3 shows the mesh occurred and the element selection for the above-mentioned problem.

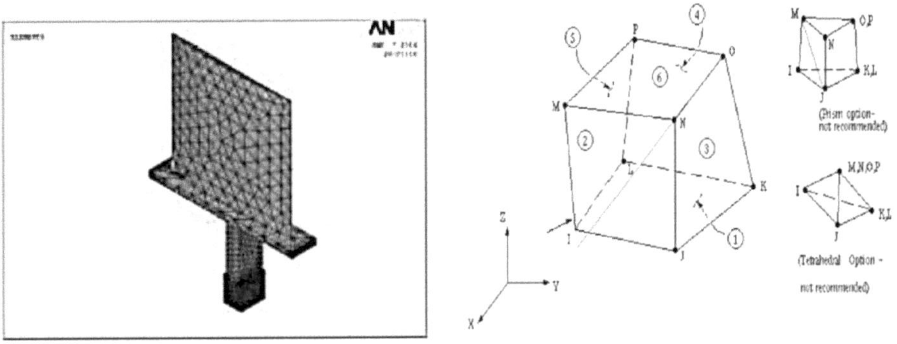

FIGURE 21.3 Finite element model of turbine blade and elements used.

Turbine blade is modeled with Solid 185 elements. Solid structure modeling in three dimensions is done with SOLID185. The element consists of eight nodes having three degrees of freedom defining the complete three-dimensional element. The complete convergence test has been followed for selecting the above element which can sustain the excess loads and deflection.

21.4 MODELLING AND MESHING

Figure 21.4 shows the modeling of the three-dimensional structure of a turbine blade with a different extrusion method.

The design of the turbine blade is carried out in ANSYS, for which modeling is being followed with respect to the defined dimensions. The blade is formed by extruding the area as per the mentioned dimensions. The turbine blade is fixed to a base which will be attached to the turbine shaft. The turbine blade was modeled with NACA 63203 airfoil, which is a widely used airfoil cross-section for turbine blades.

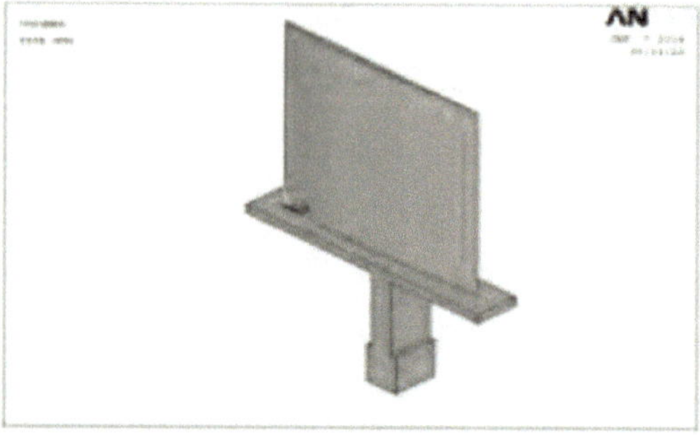

FIGURE 21.4 Turbine blade.

21.5 RESULTS

The following contour plots are being procured for INCONEL 625 ALLOY. Figure 21.5 shows the thermal gradient and deflection along three axes. The gradient shows deformation gradient with different colors emphasizing the impact of highest and lowest deformation in red and blue color.

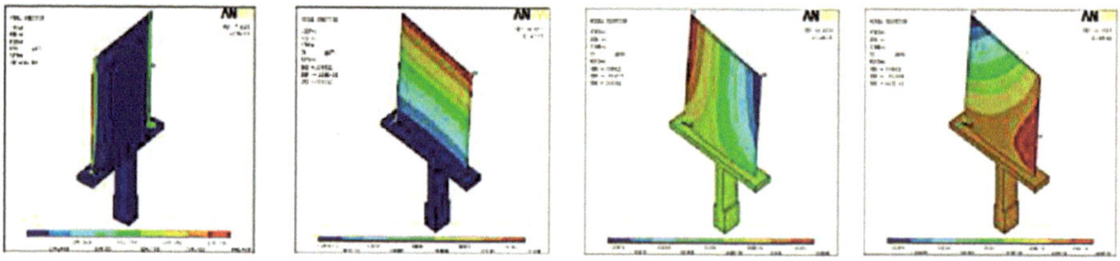

FIGURE 21.5 Thermal gradient and deformation along X, Y, and Z-axis.

Figure 21.6 shows the stress along three directions by contour plot for identification of maximum and minimum stress.

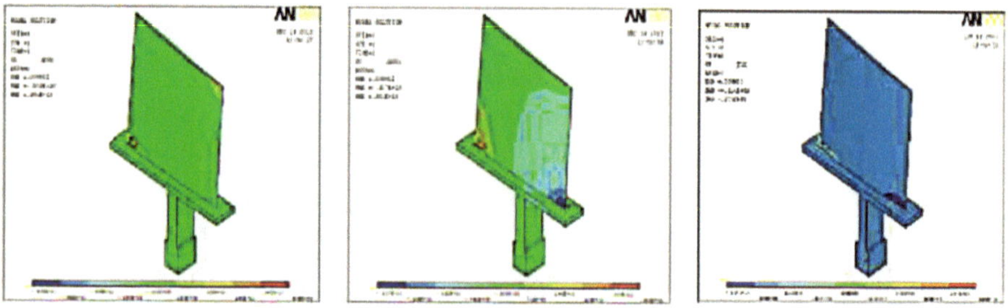

FIGURE 21.6 Stress along X, Y, and Z-direction.

Figures 21.7 and 21.8 show the results for N155 material used in turbine blades for the sake of improving the deformation properties of the turbine blade. Also, the mesh shown in the figure gives the exact behavior of the blade transformation.

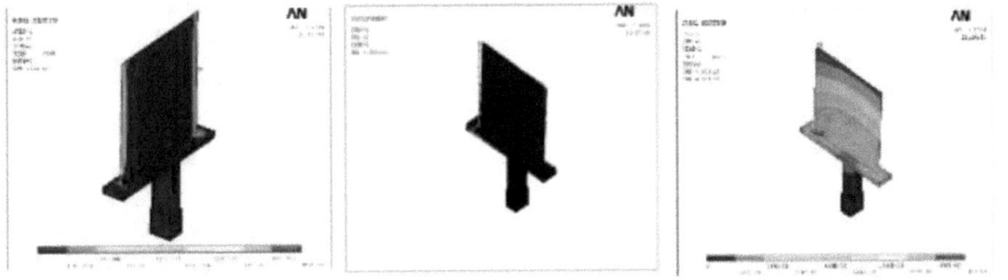

FIGURE 21.7 Nodal temperatures, deformed shape, and displacement vector sum.

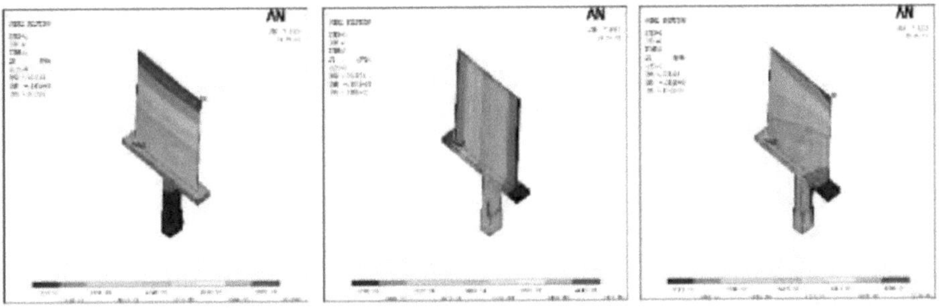

FIGURE 21.8 Deformation along X-direction.

Figure 21.9 shows the usage of material Hastelloy X for the turbine blade, with changes in deformation properties. As the contour plot shows, there is a range of deformation extremities, with red indicating the highest deformation and blue indicating the least deformation. The graphs in Figure 21.9 show the deformation along the X, Y, and Z directions.

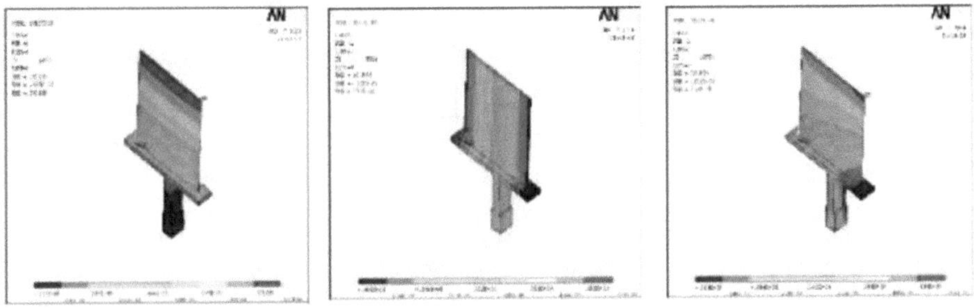

FIGURE 21.9 Deformation along X, Y, and Z-direction.

21.6 CONCLUSION

The turbine blade is allowed to expand freely in all directions. The pressure is applied at the tip of the turbine blade and at the leading edge of the blade. Thermal analysis was carried out to know the thermal stresses such as the temperature distribution, thermal gradients, and thermal fluxes of the gas turbine rotor blades. Significant results have come from analyzing heat exchangers, turbines, internal combustion engines, and piping systems. Heat flux equal to zero was applied to the areas of the base and top of the blade. Both sides of the base of the blade that come into contact with similar areas were assumed to be insulated. In the convection boundary condition, the convection heat transfer coefficient and the temperature of the surrounding gases (T) have to be specified for the areas subjected to convection.

KEYWORDS

- **blades**
- **failures**
- **finite element method**
- **nodal temperatures**
- **thermal fluxes**
- **thermal gradients**
- **turbine**

REFERENCES

1. Ghosh, S., Kundu, B., Ghosh, D., Saha, K., Chatterjee, I., & Ghosh, S., (2022). Thermal-hydraulic analysis of axial gas turbine blade. *Materials Today: Proceedings*.
2. Gill, H. S., (2021). Review of structural and thermal analysis of gas turbine blade. *Materials Today: Proceedings*, *37*, 3697–3699.
3. Moghanlou, F. S., Vajdi, M., Motallebzadeh, A., Sha, J., Shokouhimehr, M., & Asl, M. S., (2019). Numerical analyses of heat transfer and thermal stress in a ZrB_2 gas turbine stator blade. *Ceramics International*, *45*(14), 17742–17750.
4. Vaferi, K., Nekahi, S., Vajdi, M., Moghanlou, F. S., Shokouhimehr, M., Motallebzadeh, A., Sha, J., & Asl, M. S., (2019). Heat transfer, thermal stress, and failure analyses in a TiB_2 gas turbine stator blade. *Ceramics International*, *45*(15),19331–19339.

5. Rani, S., Agrawal, A. K., & Rastogi, V., (2017). Failure analysis of a first stage IN738 gas turbine blade tip cracking in a thermal power plant. *Case Studies in Engineering Failure Analysis*, *8*, 1–10.
6. Ahmed, S. I., Balguri, P. K., Govardhan, D., & Habeeb, S., (2021). Numerical study and heat transfer analysis of nickel-chromium and titanium carbide gas turbine blades cooling. *Materials Today: Proceedings*, *17*.
7. Yang, T., Xue, S., Zheng, L., Liu, L., & Liu, X., (2021). Crack investigation of martensitic stainless steel turbine blade in thermal power plant. *Engineering Failure Analysis*, *127*, 105553.

CHAPTER 22

SIMULATION OF THREE PHASE INDUCTION MOTOR SPEED CONTROL USING VOLTAGE SOURCE INVERTER

ALBERT JOHN VARGHESE,[1] SHADAB MANSHURI,[2] SAMEER AHMED,[2] SAPNA ARYA,[2] and SANDEEP KUMAR[2]

[1]*Assistant Professor, Electrical Engineering, RCET, Bhilai, Chhattisgarh, India*

[2]*UG Student, Electrical Engineering, RCET, Bhilai, Chhattisgarh, India*

ABSTRACT

The proposed study is an investigation into the modernization of speed control methods for three-phase induction motors with an aim to overcome the limitations of traditional techniques. Industries have relied on DC motors traditionally for variable speed applications due to their ease of speed control and high starting torque, despite the high maintenance costs. However, advancements in power electronics have enabled induction motors to be viable options for variable speed applications as well. The study employs MATLAB/Simulink to model and simulate a system utilizing voltage source inverters (VSI) for speed control. Through the implementation of proportional-integral-derivative (PID) controllers and voltage-frequency (V/f) control schemes, the study demonstrates successful speed regulation of the induction motor. Various response curves and measurements, such as rotor speed and stator voltage, are analyzed to evaluate the system's performance under different operating conditions. The results indicate that the V/f

Computational Optimization, Modeling, and Simulation for Engineering Applications.
Anupam Shukla, Sourabh Rungta, Mohan Awasthy, and Rakesh Himte (Eds.)

control strategy, integrated with VSI and PID control, effectively regulates motor speed. This approach showcases the potential of power electronics in enhancing industrial processes, providing insights into alternative speed control methodologies beyond traditional DC motors.

22.1 INTRODUCTION

Our country gained independence in 1947, but even after many decades, our industries still use traditional machinery and control methods. This is true not only for small industries but also for multinational corporations. Traditionally, DC motors are mostly used because their speed control is easy, and DC series motors provide high starting torque, which helps in moving heavy weight machines easily. However, the maintenance cost of DC motors is too high. Induction motors are also used, and they have lower maintenance costs. Traditionally, it is accepted that DC motors are preferred for variable speed applications, while induction motors are preferred for fixed speed applications. However, the advent of power electronics has made a drastic change, and nowadays, induction motors can also be used for variable speed applications.

An induction motor uses an alternating electric current in the stator which produces an alternating magnetic field. The magnetic field of the stator winding provides the electromagnetic induction required to produce torque in the rotor [1]. As a result, an induction motor without any electrical connections to the rotor can be constructed. An induction motor's rotor could be:

- Wound rotor induction motor; or
- Squirrel cage induction motor.

Because they are self-starting, dependable, and cost-effective, induction motors with three phases are known as squirrel-cage induction motors. Commonly employed as industrial drives. Induction motors with only one phase are known as single-phase induction motors and are frequently used for lesser loads, such as fans in the home. Induction motors with squirrel cages are frequently utilized in applications for fixed-speed and variable-frequency drives. Traditional methods used for speed control of induction motors are illustrated in brief [2]:

1. **Pole Changing:** Because the number of poles can rotate at a synchronous speed that is inversely proportional to the number of

poles of a squirrel cage does not have any wounds. A set the total number of poles. The amount of poles produced is the same as with stator winding. As a result, a squirrel cage motor merely requires an arrangement to change the number of poles in the stator [2].

2. **Stator Voltage Control:** By lowering the stator voltage, the speed of a high-slip induction motor can be decreased to a level that allows some fan and pump drives to be controlled. Current is proportional to voltage, whereas torque is proportional to voltage squared. As a result, as the voltage is dropped, the speed is reduced. Variable voltage, AC voltage controllers are used to achieve speed control [2].

$$\eta = P_m/P_g = (1 - s)(1)$$

3. **Supply Frequency Control:** The supply frequency can be changed to control the motor speed. Supply frequency multiplied by air-gap flux determines the voltage induced in the stator. If the stator drop is taken into account, terminal voltage is proportional to the product of frequency and flux [2].

$$T_{max} = K(V/f)^2/2\pi(L_s + L'_r)(2)$$

4. **Rotor Resistance Control:** The rotor resistance control speed-torque curves are shown in the diagram below. While maximum torque is present, unaffected by resistance of the rotor, the speed at which maximum torque is generated varies with rotor resistance. When rotor resistance is increased, speed decreases for the same torque. The advantage of control of rotor resistance is that it maintains motor torque capabilities even when traveling at low speeds [2].

5. **Slip Power Recovery:** An equivalent circuit of a wound-rotor induction motor with voltage Vr injected into it, assuming stator-to-rotor turns ratio unity. When rotor copper loss is neglected [2].

$$P_m = P_g - P_r(3)$$

An inverter is a piece of power electronics or circuitry that converts electricity from one form to another, such as converting direct current to alternating current at the desired frequency and voltage

[b]. This can be classified depending on the source of supply as well as the power circuit's related topology. As a result, they are split into two groups categories:

i. **Voltage Source Inverter (VSI):** It is a device that converts a DC input voltage to its AC counterpart at the output. Because the DC source at the input has a voltage, it's also known as a voltage-fed inverter (VFI), very low or no resistance. An ideal VSI keeps the voltage constant throughout the process.

ii. **Current Source Inverter (CSI):** A current inverter for the source is an inverter circuit that converts DC current into equivalent AC current at its input. It is sometimes referred to as a Current Fed Inverter and is abbreviated as CSI (CFI). The input current is kept constant in this case. A DC voltage source with a high impedance feeds CSI with an adjustable current source.

22.2 METHODOLOGY

Simulink is a block diagram environment for multi-domain simulation and Model-Based Design. Simulink is a programming environment with a graphical user interface. Based on MATLAB that may be used to model, simulate, and analyze dynamical systems with several domains. Its main interface consists of a graphical block diagramming tool and a set of block libraries that can be customized. Simulink is frequently utilized in model-based and multidomain design for automatic digital signal processing and control [3].

The software SimscapeElectricalTM Specialized Power Systems is developed to simulate power electronic devices. As an example: In the Power Electronics library, Simscape > Electrical > Specialized Power Systems> Simscape>Electrical>Specialized Power Systems > Simscape > Electrical > Special adds an IGBT/diode block. Open the thyristor block parameter window and specify the following values – Table 22.1 provides the IGBT specifications. The snubber circuit is a vital part of the thyristor dialog box.

TABLE 22.1 Parameters of IGBT

Power Electronic Devices	IGBT/Diode
Internal resistance	1e-3 Ω
Snubber resistance	1e-5 Ω
Snubber capacitance	Inf

The block diagram of Induction motor speed control using VSI is shown in Figure 22.1.

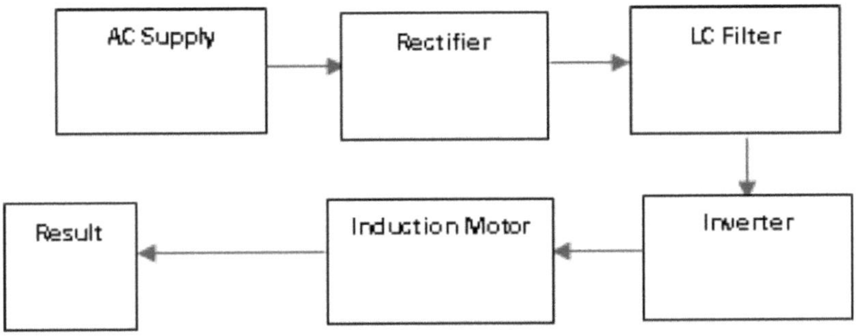

FIGURE 22.1 VSI-based induction motor speed control block diagram. [AC Supply: three-phase pure balanced AC source; Rectifier: it converts AC to DC and feeds it into a DC connection; DC link: it is used to convert pulsating DC into smooth DC; Inverter: voltage source inverter it is used to convert DC into AC; and Induction Motor: it is an electrical motor which runs with the help of electromagnetic induction].

Some of the major blocks used in the simulation are highlighted below. Induction wound rotor machine, squirrel cage machine, and other three-phase asynchronous engines are represented by machine blocks. The motor or generator will be the two operations of this machine. The torque value from the machine determines this function:

- The machine will act as a motor when Tm is positive;
- The machine will act as a generator if Tm is negative.

The machine component's instrument value is measured and shown in Table 22.2.

TABLE 22.2 Parameters for Asynchronous/Induction Motors

Nominal power, voltage (line-line), frequency.	750 VA, 400 V, 50 Hz
Stator resistance and inductance.	0.03552 Ω, 0.000335 H
Rotor resistance and inductance.	0.02092 Ω, 0.000335 H
Mutual induction.	0.02092 H
Inertia, friction factor, pole pairs.	1.25, 0.03914, 2

A feedback-based control loop is the proportional-integral-derivative controller (PID controller or three-term controller). Frequently used in industrial control systems and other applications requiring constantly modulated control [c]. The P controller helps in stabilizing gain, but it produces a constant steady state error, while the I controller eliminates mistakes in the steady state and the D controller reduces the rate of change of error. These three actions are inherent in a PID controller. These are considered to be the most accurate and stable controller. The PID controller component's instrument value is monitored and shown in Table 22.3.

TABLE 22.3 PID Controller Parameter

Proportional	0.0001
Integral	0.09
Upper saturation limit	50
Lower saturation limit	–50

After successfully converting AC to DC, then output DC is fed into VSI separately by taking 440 V DC. Then we converted it into AC successfully. In the inverter, IGBT is used and triggered through gate pulse for operation. Then this converted AC is fed into the three-phase induction motor. After that, the motor also runs. The inverter component's instrument value is measured and shown in Table 22.4.

TABLE 22.4 Inverter Component Value of a Block Measurement

Power electronic devices	IGBT/diode
Internal resistance	1e-3 Ω
Snubber resistance	1e-5 Ω
Snubber capacitance	Inf

The instrument value of rectifier component is measured and shown in Table 22.5.

TABLE 22.5 Rectifier Component Block Measurement Value

Power Electronic Devices	Thyristor
Resistance	0.001 Ω
Inductance	Ie-3 H
Forward voltage	0.8 V
Latching current	0.1 A
Turn off time	100e-6 S
Initial current	0 A
Snubber resistance	500 Ω
Snubber capacitance	250e–9 H

Three-phase fully controlled rectifier circuit shown below. Here we have taken a three-phase balanced 440 V source and thyristor, by which the output is controlled. A DC output pulse generator is used to trigger the thyristor. An ammeter and voltmeter are used for measurement, and a scope is used for output wave.

In the intermediate circuit of several types of converters, an intermediate circuit (DC-link) separates electrical grids to one DC voltage level. After running the motor successfully through the inverter, a rectifier is connected using a DC link. Then the output is fed back to the circuit and it runs successfully.

After implementing the use of a VSI to drive MATLAB/Simulink, it was used to model a three-phase induction motor. The model was run successfully. The graphs could be generated using a scope. The inverter and rectifier sections designed worked properly and gave the proper inputs to the induction motor with three phases. The proposed system forms a closed loop, which is simulated successfully in the MATLAB/Simulink Environment. The PID control helped in maintaining the speed of the developed system. It was also noted that the V/f speed control scheme proved to be useful, and the simulation was successfully executed. Figure 22.2 shows the simulation structure of controlling the speed of a three-phase induction motor VSI.

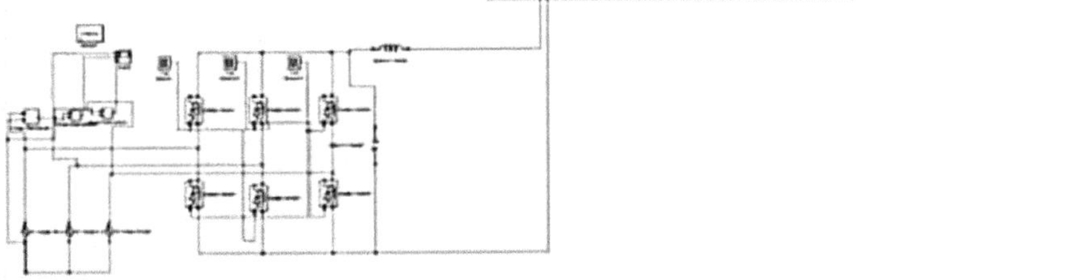

FIGURE 22.2 Circuit diagram of speed control VSI based induction motor.

22.3 RESULT AND DISCUSSION

The various response curves obtained during the execution of the Simulink model are shown below. The rotor measurements, rotor speed measurement, stator voltage measurement done using scope are plotted as graphs. This VFD can run at a variety of frequencies and voltages. Motor speed and torque for a 50 Hz f = 50 Hz drive (Figure 22.3).

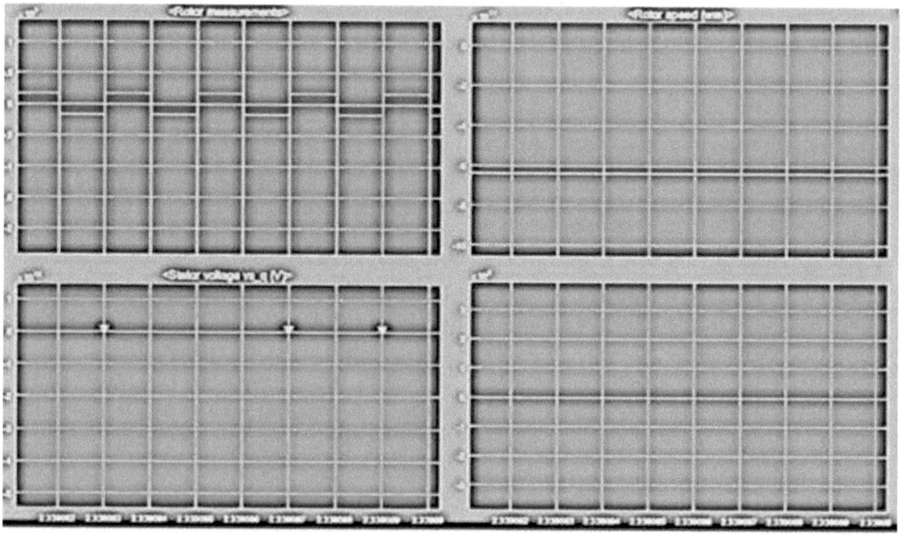

FIGURE 22.3 Rotor measurement and rotor speed static.

Observation of rotor speed and stator voltage at 440 V is shown in Table 22.6.

TABLE 22.6 Rotor Speed and Stator Voltage at 440 V

Parameters	Minimum	Maximum	RMS Value
Rotor measurement	–547.1	542.5	527.6
Rotor speed	–265.6	0	182.5
Electromagnetic torque	–388.8	827.6	497.2
Stator voltage	–314	486.1	644.1

Observation of rotor speed and stator voltage at 640 V is shown in Table 22.7.

TABLE 22.7 Rotor Speed and Stator Voltage at 640 V

Parameters	Minimum	Maximum	RMS Value
Rotor measurement	–352.8	363.3	397.7
Rotor speed	–455.3	00	315.8
Electromagnetic torque	–673	134.3	–
Stator voltage	–593.4	944.7	122

Observation of Rotor speed and Stator voltage at 720 V is shown in Table 22.8.

TABLE 22.8 Rotor Speed and Stator Voltage at 720 V

Parameters	Minimum	Maximum	RMS Value
Rotor measurement	–406.3	417	459.6
Rotor speed	–733.3	00	358.6
Electromagnetic torque	–797.6	154.7	771.9
Stator voltage	–355.3	231.5	302.9

Observation of Rotor speed and Stator voltage at 1,040 V is shown in Table 22.9.

TABLE 22.9 Rotor Speed and Stator Voltage at 1,040 V

Parameters	Minimum	Maximum	RMS Value
Rotor measurement	–932.3	946.3	607.2
Rotor speed	–328.3	00	212.8
Electromagnetic torque	–	–	–
Stator voltage	–127.3	153.1	653.6

22.4 CONCLUSION

The using v/f control to control the speed of a three-phase induction motor is successfully implemented using MATLAB/Simulink. The VSI combined with the PID controller helped in changing the firing pulse and displayed good speed control characteristics. The power electronics-based approach for the speed control was successfully demonstrated using the proposed method. MATLAB/Simulink helped in understanding the concepts involved in the approach which is different as compared to the traditional approaches in speed control.

KEYWORDS

- **alternating current**
- **current source inverter**
- **direct current**
- **proportional integral derivative**
- **voltage by frequency**
- **voltage source inverter**

REFERENCES

1. IEC 60050 (Publication date: 1990-10). *Section 411-31: Rotation Machinery – General, IEV Ref. 411–31–10: Induction Machine – an Asynchronous Machine of Which Only One Winding is Energized.* http://www.electropedia.org/iev/iev.nsf/display?openform&ievref=411-31-10 (accessed on 21 December 2023).
2. Dubey, G. K., (2001). *Fundamentals of Electrical Drives* (2ndedn.).
3. *System Requirements and Platform Availability by Product,* (2015). Mathworks.com (accessed on 21 December 2023).
4. Hartono, H., Sudjoko, R. I, & Iswahyudi, P., (2019). Speed Control of Three-Phase Induction Motor Using Universal Bridge and PID Controller. *Journal of Physics: Conference Series, The 1st International Conference on Engineering and Applied Science 21 August 2019, Madiun, Indonesia. 1381*(012053), 1–8, DOI: 10.1088/1742-6596/1381/1/012053.

CHAPTER 23

SOLID STATE PROCESSES FOR FABRICATION OF ALUMINUM METAL MATRIX COMPOSITES FOR AUTOMOTIVE AND AIRCRAFT APPLICATIONS

AGNIVESH KUMAR SINHA,[1] KASI RAJA RAO,[1] VINAY KUMAR SONI,[2] NITIN UPADHYAY,[3] RITYUJ SINGH PARIHAR,[4] and MOHIT KUMAR SAHU[5]

[1]*Mechanical Engineering Department, Rungta College of Engineering and Technology, Bhilai, Chhattisgarh, India*

[2]*Mechanical Engineering Department, Columbia Institute of Engineering and Technology, Raipur, Chhattisgarh, India*

[3]*Mechanical Engineering Department, Madhav Institute of Technology and Science, Gwalior, Madhya Pradesh, India*

[4]*Mechanical Engineering, Shri Shankaracharya Institute of Professional Management and Technology, Raipur, Chhattisgarh, India*

[5]*Mechanical Engineering Department, ITM University, Raipur, Chhattisgarh, India*

ABSTRACT

Aluminum Metal Matrix Composites (Al-MMCs) are popular for their applications in Automotive and Aircraft applications majorly due to their high strength to weight ratio. This facilitates in achieving the light weight of

Computational Optimization, Modeling, and Simulation for Engineering Applications.
Anupam Shukla, Sourabh Rungta, Mohan Awasthy, and Rakesh Himte (Eds.)

aircraft or automotive components which in turn reduces the fuel consumption and improves the performance. This chapter deals with the various solid-state processing techniques of Al-MMCs for Automotive and Aircraft applications. This also elucidates the major research gaps identified from the literature review.

23.1 INTRODUCTION

Composite is a multi-phase material, generally consisting of two phases, matrix and reinforcement. Constituent phases of composite must be chemically dissimilar and physically separable. Constituent materials of composite should be selected such that when combined, better desired properties are realized. Evolution of composites is driven by desired properties. This evolution is due to their remarkable capability of combining materials effectively to deliver better properties for a wide range of applications. In general, composites possess a remarkable combination of mechanical properties like tensile strength, flexural strength, impact strength, and stiffness.

Matrix phase of composite is continuous phase which binds the reinforcement. Its main purpose is to transmit the load to the reinforcement for which it should have a lower value of elastic modulus than the reinforcement. Whereas reinforcement strengthens the composites and bears the load transmitted by the matrix. Properties of the composite majorly depend on the properties of the matrix material. Composites are broadly classified into three categories namely, metal matrix composites (MMCs), polymer matrix composites (PMCs), and ceramic matrix composites (CMCs).

MMC utilizes a ductile metal as a matrix. MMC can be designed for high-temperature applications despite the fact that its base metal counterpart is not suitable for the same application. The addition of reinforcement in MMCs improves specific strength, thermal conductivity, creep resistance, and resistance to abrasion. MMCs also provide service at moderate temperatures and are non-flammable. Alloys of Magnesium, Aluminum, Titanium, and Copper are some of the commonly used matrices in MMCs. Reinforcements may be fibrous, whiskers, or particulates with a volume fraction ranging from 10% to 60%. MMCs have found their application in automotive and aerospace industries besides structural, biomedical, electronics, and electrical industries. Some automobile engines, drive shafts, and transmission components utilize MMCs for their lightweight, high strength, and wear resistance attributes. Reinforcements like refractory metal wires (Ta, Mo, Nb, W), oxide fibers (Al_2O_3, SiO_2, ZrO_2), Boron fibers, SiC fibers, Carbon fibers are used for aerospace industrial applications [1].

Automotive applications of MMCs include their use in piston rods, bracing systems, pistons, frames, valve spring cap, piston pins, disc brakes, calipers, pads of brake, carbon shaft and accumulator plates. For these applications MMCs must have high specific strength and stiffness, low thermal expansion coefficient, high temperature wear resistance and creep resistance. Likewise, MMCs with properties like high stiffness and specific strength, thermal resistance and low thermal expansion coefficient are utilized for aerospace applications like joining elements, frames, reinforcements, and aerials, turbine blades, fan and compressor blades, gearbox, axle-tubes [2].

Researchers and scientists have been exploring the combinations of reinforcement in various MMCs to achieve desired properties for aerospace and automotive applications. MMCs based on Aluminum, Copper, Magnesium, and Titanium matrices have always been popular MMC alternatives among researchers specifically for automotive and aerospace applications. However, Aluminum-based MMCs have been the most researched MMCs for the aerospace and aircraft applications due to their lightweight, high corrosion resistance, high specific strength, low density, and high wear resistance attributes.

Therefore, this chapter deals with fabrication of Aluminum metal matrix composites (Al-MMCs) by solid state processes for automotive and aircraft applications where various types of Al MMCs used for automotive and aircraft applications are discussed. Solid state processing techniques used for fabrication of Al-MMCs are also elucidated. Moreover, challenges and future scope of solid-state processing techniques for fabrication of Al-MMCs for automotive and aircraft applications are also briefly discussed which would be lucrative for the scientists and researchers working for the advancement in this pertinent domain.

23.2 AL-MMCS FOR AUTOMOTIVE AND AIRCRAFT APPLICATIONS

Specifically for automotive and aerospace applications, lightweight and yet strong components are essential requirements. For this purpose, Al-MMC is one of the best alternatives among all other MMCs. Pure aluminum is inherently light, malleable, and ductile. Various reinforcements in different forms are used to strengthen Al-MMCs. Commonly used reinforcements are

SiC, Al_2O_3, B_4C, TiB_2, TiC, TiO_2, etc. Likewise, aluminum and its alloys are also used as matrices in the fabrication of Al-MMCs. Aluminum is alloyed with different elements like Cu, Si, Zn, Mg, Mn, and a combination of Mg and Si to form aluminum alloys designated as AA6000, AA4000, AA7000, AA5000, AA3000, and AA6000 series, respectively. Wherein, AA5000 and AA3000 series are non-heat treatable and the remaining other alloys are heat treatable.

For the application of composite components in the automotive and aircraft industries, workability is an essential requirement. Researchers investigated the forgeability of Al-MMCs [3] as it determines the workability of composite materials. SiC was used as reinforcement in the aluminum matrix. The weight percentage of SiC was varied from 0% to 20% in step sizes of 5%. Furthermore, the hardness of various composites was also determined. The study showed that an increase in weight percentage of SiC reduces the forgeability of MMCs. It was also observed that hardness increased with increase in wt.% of SiC. Another study on Al-MMCs manifested that addition of SiC and B_4C resulted in increase in modulus of elasticity of 6,061 aluminum alloy [4]. Results also showed that Al-MMC with a large quantity of carbides tends to have better mechanical properties (like hardness, tensile, flexural strength). Researchers have also investigated the influence of SiC particles on Al-MMCs having 40% copper [5]. In step size of 2.5%, SiC was reinforced in the range of 0% to 15% which led to the reduction in thermal conductivity of Al-MMCs whereas, compressive strength increased. Addition of SiC first led to deterioration in wear rate of Al-MMCs up to 7.5 wt.% and then increased. High wear resistance of Al-MMCs could be utilized in disc brake components in automotive and aerospace applications.

Functionally graded components for automotive applications were also developed based on pure Al/Al_2O_3 MMCs [6]. Both wear resistance and mechanical properties of composites improved with incorporation of Al_2O_3 particles. Besides mechanical properties and wear resistance, corrosion performance of Al-MMCs was also improved with the addition of Ti-6Al-4V particles in 7075 aluminum alloy matrices for their application in automotive components like disc brakes, engine cylinders, cardan shafts, pistons, drum brakes, etc. [7]. Hardness, tensile strength, and wear resistance of Al-MMC were improved by the addition of 5 wt.% TiC [8]. Moreover, researchers have also utilized ZrO_2 particles and nano-graphite together in 7,075 aluminum alloy to improve impact strength, tensile strength, hardness, and wear resistance of composites [9].

From literature, it is very evident that ceramic particulates are often used for improving mechanical and tribological attributes of Al-MMCs for their aircraft and automotive applications. However, studies also suggest that incorporation of solid lubricants such as MoS_2 and graphite [10] in combination with ceramic particulates (nano- or micro-sized) could drastically enhance the tribological and mechanical properties of aluminum MMCs.

23.3 SOLID STATE PROCESSING OF AL-MMCS FOR AUTOMOTIVE AND AIRCRAFT APPLICATIONS

Production methods for Al-MMCs have evolved over time to take into account the simplicity with which they may be manufactured and the ultimate quality of the desired composite material. In accordance with the findings of the research, Al-MMCs of the appropriate size and shape have been manufactured using both liquid-state and solid-state fabrication methods. Although there are many potential challenges to obtaining Al-MMCs that are defect-free, the most common are improper selection of processing and production procedures causing difficulties [11–17]. Furthermore, in this section, the solid-state manufacturing technique of Al-MMCs, as well as their distinct characteristics for automotive and aerospace applications, have been thoroughly explored. On the basis of their study findings, they have also developed a clear explanation of solid-state processing factors and result characteristics that are detailed.

23.3.1 POWDER METALLURGY

In solid-state manufacturing techniques, reinforcing substance is injected into the matrix body and incorporated into the matrix's anatomy as a whole while the matrix is still in the solid state [18]. One of the solid-state manufacturing techniques is high energy planetary ball milling, which is a powder processing technique that makes use of high-frequency hard material balls with an attrition mill for mechanical milling of processed powders. Powder flecks become trapped across dynamic impacting balls, causing repeated distortion, re-welding, and dissociation of blended powder particles, which results in homogeneous distribution of particles in the aggregate. Ceramic hard particles are incorporated into metal alloys to form composites that have increased strength, hardness, and wear resistance [19]. To prevent powder from reacting with the surrounding environment during this operation, a controlled atmosphere is desirable. It is a cumbersome process that is unsuitable for large-scale manufacturing.

An aluminum metal matrix reinforced with 15 wt.% SiC and 7 wt.% TiC particles was produced by researchers [20] using microwave and traditional sintering techniques. The alloys used were pure aluminum and 1056 aluminum, with SiC and TiC particles as reinforcement. Microwave sintering of MMCs resulted in a considerable enhancement in the mechanical properties of composites, according to the findings. When sintered at 750°C, the highest bending strength was achieved, which was around 340 MPa, and the Vickers hardness was approximately 192 HV. These significant changes can be attributed to the homogeneous particle dispersion of SiC and TiC.

Research [21] reported the use of powder metallurgy techniques in the development of Al-Cu based MMCs reinforced with SiC particles, which might be a candidate material for aerospace applications. It has been explained how extrusion, alongside the volume fraction of filled particles, affects the final product. As a result of the extrusion process, an increase in the homogeneity of SiC particle dispersion in the Al matrix was observed, which improved the interfacial bonding strength of the composite. Another research on the production of Al-MMCs with SiC particles using powder metallurgy [22] stated that the aluminum matrix composite, in comparison with unreinforced aluminum alloy, exhibited considerably higher elastic modulus, yield strength, and tensile strength. The uniform dispersion of SiC particles was responsible for this improvement. Using powder metallurgy, researchers have [23] produced an aluminum matrix composite strengthened with SiC particles. The homogeneous circulation of SiC particles within the matrix improved the grain of the aggregate, and dislocation strengthening were the factors that contributed to the improvement of the mechanical properties of the composite. Figure 23.1 manifests high-density dislocation in the composite.

In another research [24], thermal conductance of Al-MMCs reinforced with SiC particles was manufactured using ball mill and plasma spraying methods were determined. It was discovered that the thermal conductance of MMCs decreased at ambient temperature. It was due to the prolonged agitation time and micron size of reinforcement particles used.

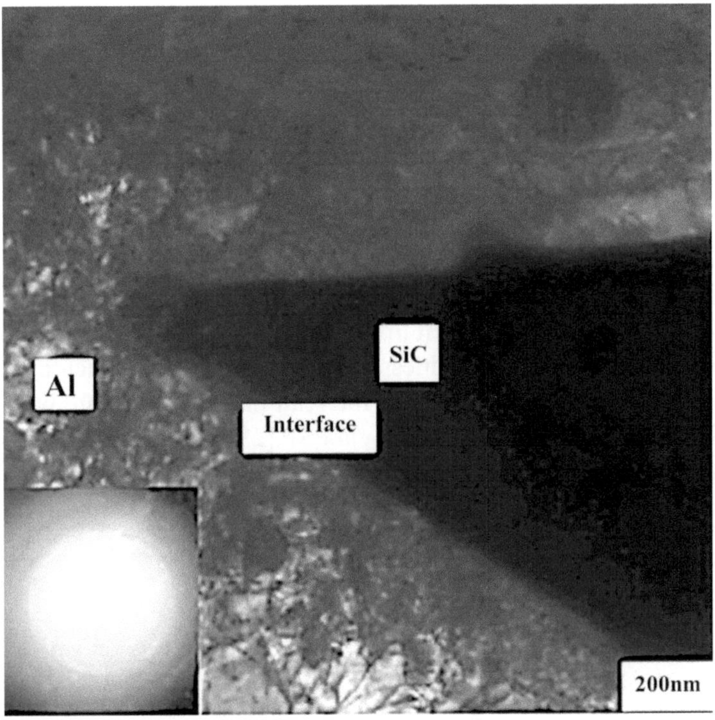

FIGURE 23.1 TEM image of SiC/Al MMC showing high density of dislocations in matrix.
Source: Reprinted with permission from Ref. [23]

23.3.2 FRICTION STIR PROCESSING (FSP)

FSP is often used to modify the microstructure, texture, and material property locally by plastic deformation. It is one of the widely used techniques for fabrication of high-performance Al-MMCs for automotive and aerospace applications. Aluminum alloy AA6061 was reinforced with fly ash particles via FSP in research [25], which manifested homogeneous distribution of fly ash. The study also reported grain refinement of MMCs, which was attributed to the pinning effect of fly ash and dynamic recrystallization. Addition of fly ash also improved microhardness of MMCs.

Process parameters such as tool rotation speed, number of passes, and traverse speed were analyzed along with the influence of SiC particle size on the microstructure and properties of Al-MMCs [26]. The study showed that the incorporation of SiC, an increase in the number of passes, and a decrease in the particle size of SiC resulted in improving hardness as well as wear resistance of 5052AA/SiC MMCs. Researchers have also fabricated

Al-MMCs reinforced with tungsten particles via FSP [27]. The results showed that an increase in the number of passes improved the micro-hardness of composites. It was also observed that the incorporation of W particles improved the electrical conductivity of the composite. However, the number of passes had no influence on electrical resistivity.

Surface-hybrid MMC layers on aluminum plate was also developed using FSP for which SiC and Al_2O_3 particles were reinforced in aluminum [28]. After the first pass, reinforcing particles (mainly SiC) were arranged in a band-like structure which was distinctly visible in the scanning electron microscope (SEM) image taken in the vicinity of the nugget region. Results showed that the average hardness deteriorated with an increase in Al_2O_3 particles in MMCs. The surface of the Aluminum alloy plate was modified with the dispersion of SiC particles via FSP [29], which resulted in excellent bonding between the base metal plate and SiC, and a uniform distribution of SiC in the MMC as shown in Figure 23.2.

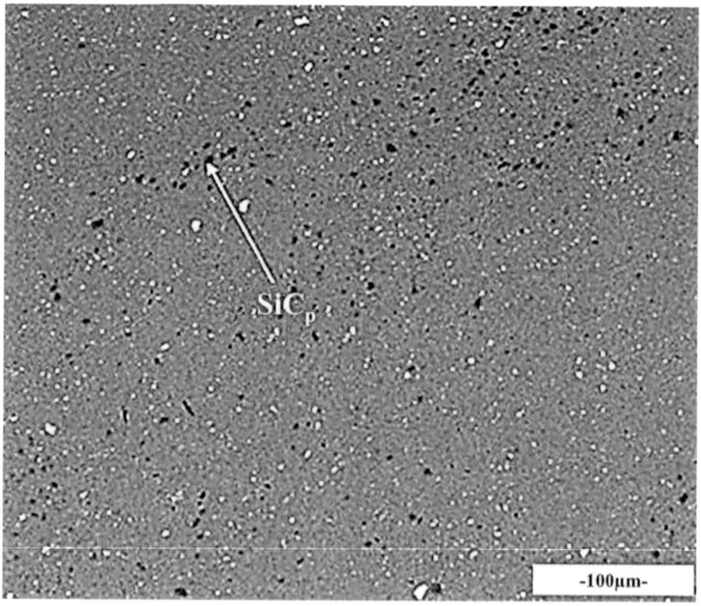

FIGURE 23.2 SEM backscattered electron image of SiC/5A06 composite.

Source: Reprinted with permission from Ref. [29]

23.3.3 DIFFUSION BONDING

This technique of fabrication of MMCs involves the application of pressure and heat for consolidation in which atoms diffuse together in the solid state.

Diffusion bonding is preferred for its advantages like the prevention of cracks and grain boundary distortion [30]. In this process, the temperature of about 50 to 70% of the melting point is maintained for the optimum time required to develop bonds in the interface region. Temperature, dwell time, and pressure are the three significant process parameters that govern the properties of MMCs. All these three parameters must be optimum to ensure desired properties. It is a simple and easy process. However, the cost of removing the oxide layer is high, which limits its application for the development of Al-MMCs for automotive applications. Despite this issue, researchers have made several attempts to develop Al-MMCs via diffusion bonding. One such research showed that Al-MMCs fabricated by vacuum-free diffusion bonding technique exhibited a clean joining surface with an average joint efficiency of 90.5% [31].

Most commonly used Al-MMCs in automotive and aerospace applications are Al/SiC MMCs. Researchers have also carried out investigations on the vacuum diffusion bonding attributes of 2024 Al/SiC [32]. It involved the study of joining attributes of similar and dissimilar MMCs and the effect of SiC particulates on bonding characteristics and their properties. Results showed that bond strength decreases with an increase in the volume percentage of SiC particulates in Al-MMCs. The study also suggested that the use of appropriate soft insert alloy coatings or interlayers might result in enhancing the quality of the joint. Similarly, the characteristics of diffusion bonded pure Al/SiC MMCs were also determined in the research [33].

In another research, the influence of diffusion bonding process parameters on the properties of Al-MMCs was determined [34]. In this study, Aluminum foil (Al-0.42Fe-0.13Si) was used as the matrix, which was reinforced with 304 stainless steel fibers. It was observed that the tensile strength of MMCs decreases with an increase in hot pressing time (5–60 min) and temperature (475°C–600°C). However, the yield strength increased with an increase in temperature as shown in Figure 23.3.

23.4 FUTURE DIRECTIONS

This chapter presents various solid-state processing techniques for fabrication of Al-MMCs for automotive and aerospace applications. Each of the techniques has its own advantages. However, the use of solid-state processing routes for the development of Al-MMCs is still limited and has a long way ahead. There are several issues that need to be addressed pertinent to the fabrication of Al-MMCs for their applications in automotive and aerospace industries.

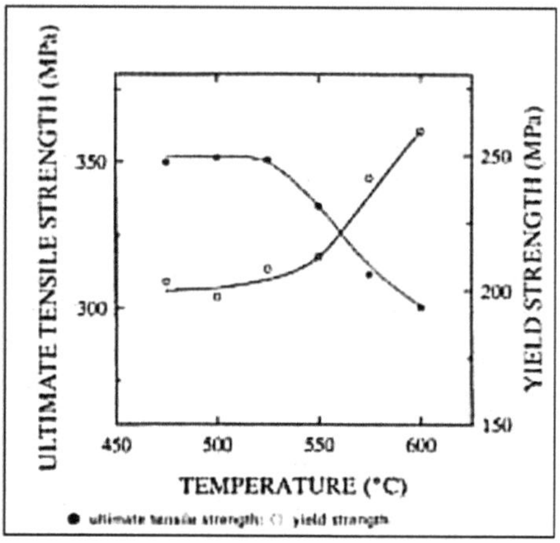

FIGURE 23.3 Tensile properties of Al-MMCs when hot pressed at 70 MPa and 30 min.
Source: Reprinted with permission from Ref. [34]

Literature shows that distribution of reinforcements, porosity, and wettability affects the mechanical properties of composites. Thus, process parameter optimization becomes necessary. Statistical modeling methods like response surface methodology (RSM) and Taguchi method can be employed for optimization of process parameters. Some of the researchers have used vacuum fabrication techniques to prevent contamination and reduce porosity. Mass production of Al-MMCs is another aspect which limits the use of MMCs in automotive and aerospace applications. All techniques are not suitable for mass production. For instance, most of the research on Al-MMC reported use of powder metallurgy for fabrication of composites. But powder metallurgy route is expensive as it requires raw materials in powder form which is generally more expensive than their bulk forms. And gas contamination is another issue which makes this route unsuitable for mass production despite its ability to produce superior performing Al-MMCs which could be used in aerospace and automotive applications. It was also observed from literature that SiC was the most used reinforcement in Al-MMCs. And there is scarcity of literature based on TiC, W, Al_2O_3, graphite, ZrO_2, SiO_2, B_4C, TiB_2, TiO_2 reinforcements in Al-MMCs. Fracture toughness is also one of the properties which is important for components used in automotive and aerospace applications. XFEM modeling and analysis of Al-MMCs could be beneficial for determination of fracture toughness and propagation of cracks.

23.5 SUMMARY AND CONCLUSIONS

In this chapter, various solid state processing techniques were discussed along with the effect of their process parameters on the properties of Al-MMCs. Major conclusions drawn from this chapter are as follows:

1. SiC, TiC, fly-ash, W-particles, Al_2O_3, stainless steel fibers, etc., were mostly utilized as reinforcement alternatives in Al-MMCs.
2. SiC particles as reinforcement in Al-MMCs is the most popular alternative among researchers.
3. Incorporation of SiC in Al-MMCs resulted in improved hardness, wear resistance, mechanical properties (like yield strength, elastic modulus, tensile strength, etc.) of composites. Moreover, addition of SiC also led to dislocation strengthening and enhanced interfacial bonding of Al-MMCs.
4. It was also observed from literature that powder metallurgy route is the most preferred solid-state processing technique for fabrication of Al-MMCs. It is due to the fact that powder metallurgy produces composites which exhibit excellent density, hardness, mechanical properties, and wear attributes. However, this route is not suitable for mass production as it is expensive.
5. Literature also reveals that there are a limited number of studies available on diffusion bonding and FSP solid-state processing techniques for the fabrication of Al-MMCs for automotive and aerospace applications.

All the processes in solid state processing techniques have a lot of potential for the development of high-performance Al-MMCs for aerospace and automotive applications, which is not yet fully explored to its limits.

KEYWORDS

- **aircraft applications**
- **aluminum metal matrix composites**
- **automotive applications**
- **diffusion bonding**
- **fabrication**
- **powder metallurgy**
- **solid state processing**

REFERENCES

1. Anthony, P. B. N., Run, R., & Watt, G. (1989). *Introduction to Metal Matrix Composites in Aerospace Applications, 1.*
2. Ulrich, K. K., (2021). Basics of metal matrix composites. In metal matrix composites, basics of metal matrix composites. In: *Metal Matrix Composites: Custom-Made Materials for Automotive and Aerospace Engineering* (pp. 1–54). Wiley-VCH, 2006. [Online]. Available: https://doi.org/10.1002/3527608117.ch1. ISBN: 3-527-31360-5.
3. Das, S., Behera, R., Datta, A., Majumdar, G., Oraon, B., & Sutradhar, G., (2010). Experimental investigation on the effect of reinforcement particles on the forgeability and the mechanical properties of aluminum metal matrix composites. *Materials Sciences and Applications, 1*(5), 310–316. doi: 10.4236/msa.2010.15045.
4. Reddy, P. S., Kesavan, R., & Vijaya, R. B., (2018). Investigation of mechanical properties of aluminum 6061-silicon carbide, boron carbide metal matrix composite. *Silicon, 10*(2), 495–502. doi: 10.1007/s12633-016-9479-8.
5. Selvakumar, V., Muruganandam, S., & Senthilkumar, N., (2017). Evaluation of mechanical and tribological behavior of Al–4%Cu–x%SiC composites prepared through powder metallurgy technique. *Transactions of the Indian Institute of Metals, 70*(5), 1305–1315. doi: 10.1007/s12666-016-0923-7.
6. Saleh, B. I., & Ahmed, M. H., (2019). Development of functionally graded tubes based on pure Al/Al_2O_3 metal matrix composites manufactured by centrifugal casting for automotive applications. *Metals and Materials International.* doi: 10.1007/s12540-019-00391-3.
7. Adetunla, A., & Akinlabi, E., (2019). Fabrication of aluminum matrix composites for automotive industry via multipass friction stir processing technique. *International Journal of Automotive Technology, 20*(6), 1079–1088. doi: 10.1007/s12239-019-0101-0.
8. Mohapatra, J., Nayak, S., & Mahapatra, M. M., (2020). Mechanical and tribology properties of Al-4.5%Cu-5%TiC metal matrix composites for light-weight structures. *International Journal of Lightweight Materials and Manufacture, 3*(2), 120–126. doi: 10.1016/j.ijlmm.2019.09.004.
9. Kumar, R., Deshpande, R. G., Gopinath, B., Harti, J., Nagaral, M., & Auradi, V., (2021). Mechanical fractography and worn surface analysis of nanographite and ZrO_2-reinforced Al7075 alloy aerospace metal composites. *Journal of Failure Analysis and Prevention, 21*(2), 525–536. doi: 10.1007/s11668-020-01092-5.
10. Bhaskar, S., Kumar, M., & Patnaik, A., (2019). Mechanical and tribological overview of ceramic particulates reinforced aluminum alloy composites. *Reviews on Advanced Materials Science* (Vol. 58, No. 1, pp. 280–294). Walter de Gruyter GmbH. doi: 10.1515/rams-2019-0033.
11. Vineeth, K. K., & Jayahari, L., (2018). Study of mechanical properties and wear behavior of aluminum 6063 matrix composites reinforced with steel machining chips. *Materials Today: Proceedings, 5*(9), 20285–20291. doi: 10.1016/j.matpr.2018.06.400.
12. Pramanik, A. (2008). *Explicit Understanding of Reinforcement Effects on Wear Resistance of Metal Matrix Composites*, 1–19.
13. Vedrtnam, A., & Kumar, A., (2017). Fabrication and wear characterization of silicon carbide and copper reinforced aluminum matrix composite. *Materials Discovery, 9*, 16–22. doi: 10.1016/j.md.2018.01.002.

14. Sethi, J., Das, S., & Das, K., (2019). Study on thermal and mechanical properties of yttrium tungstate-aluminum nitride reinforced aluminum matrix hybrid composites. *Journal of Alloys and Compounds, 774*, 848–855. doi: 10.1016/j.jallcom.2018.10.017.
15. Selvakumar, S., Dinaharan, I., Palanivel, R., & Babu, B. G., (2017). Development of stainless-steel particulate reinforced AA6082 aluminum matrix composites with enhanced ductility using friction stir processing. *Materials Science and Engineering A, 685*, 317–326. doi: 10.1016/j.msea.2017.01.022.
16. Hima, G. C., Durga, P. K. G., Ramji, K., & Vinay, P. V., (2018). Mechanical characterization of aluminum metal matrix composite reinforced with aloe vera powder. *Materials Today: Proceedings, 5*(2), 3289–3297. doi: 10.1016/j.matpr.2017.11.571.
17. Li, G. R., Cheng, J. F., Wang, H. M., & Li, C. Q., (2017). The influence of cryogenic-aging circular treatment on the microstructure and properties of aluminum matrix composites. *Journal of Alloys and Compounds, 695*, 1930–1945. doi: 10.1016/j.jallcom.2016.11.028.
18. Aynalem, G. F., (2020). Processing methods and mechanical properties of aluminum matrix composites. *Advances in Materials Science and Engineering, 2020.* doi: 10.1155/2020/3765791.
19. Sharma, D. K., Mahant, D., & Upadhyay, G., (2019). Manufacturing of metal matrix composites: A state of review. *Materials Today: Proceedings, 26*, 506–519. doi: 10.1016/j.matpr.2019.12.128.
20. Ghasali, E., Yazdani-rad, R., Asadian, K., & Ebadzadeh, T., (2017). Production of Al-SiC-TiC hybrid composites using pure and 1056 aluminum powders prepared through microwave and conventional heating methods. *Journal of Alloys and Compounds, 690*, 512–518. doi: 10.1016/j.jallcom.2016.08.145.
21. Wang, Z., Song, M., Sun, C., Xiao, D., & He, Y., (2010). Effect of extrusion and particle volume fraction on the mechanical properties of SiC reinforced Al-Cu alloy composites. *Materials Science and Engineering A, 527*(24, 25), 6537–6542. doi: 10.1016/j.msea.2010.07.017.
22. Cheng, N. P., Li, C. M., Hui, Q., & Chen, Z. Q., (2009). Effect of particle surface treatment on the microstructure and property of SiCp/AA6066 composite produced by powder metallurgy. *Materials Science and Engineering A, 517*(1, 2), 249–256. doi: 10.1016/j.msea.2009.04.005.
23. Cheng, N. P., Zeng, S. M., & Liu, Z. Y., (2008). Preparation, microstructures, and deformation behavior of SiCP/6066Al composites produced by PM route. *Journal of Materials Processing Technology, 202*(1–3), 27–40. doi: 10.1016/j.jmatprotec.2007.08.044.
24. Gui, M., Kang, S. B., & Euh, K., (2005). Thermal conductivity of Al-SiCp composites by plasma spraying. *Scripta Materialia, 52*(1), 51–56. doi: 10.1016/j.scriptamat.2004.08.028.
25. Dinaharan, I., & Akinlabi, E. T., (2018). Low-cost metal matrix composites based on aluminum, magnesium and copper reinforced with fly ash prepared using friction stir processing. *Composites Communications,9*, 22–26. doi: 10.1016/j.coco.2018.04.007.
26. Dolatkhah, A., Golbabaei, P., Besharati, G. M. K., & Molaiekiya, F., (2012). Investigating effects of process parameters on microstructural and mechanical properties of Al5052/SiC metal matrix composite fabricated via friction stir processing. *Materials and Design, 37*, 458–464. doi: 10.1016/j.matdes.2011.09.035.

27. Huang, G., Shen, Y., Guo, R., & Guan, W., (2016). Fabrication of tungsten particles reinforced aluminum matrix composites using multi-pass friction stir processing: Evaluation of microstructural, mechanical, and electrical behavior. *Materials Science and Engineering A*, *674*, 504–513. doi: 10.1016/j.msea.2016.07.124.
28. Mahmoud, E. R. I., Takahashi, M., Shibayanagi, T., & Dceuchi, K., (2009). Fabrication of surface-hybrid-MMCs layer on aluminum plate by friction stir processing and its wear characteristics. in *Materials Transactions*, *50*(7), 1824–1831. doi: 10.2320/matertrans.M2009092.
29. Wang, W., Yu, S. Q., Liu, P., Ke, L. H., & Li, T., (2009). A novel way to produce bulk SiCp reinforced aluminum metal matrix composites by friction stir processing. *Journal of Materials Processing Technology*, *209*(4), 2099–2103. doi: 10.1016/j.jmatprotec.2008.05.001.
30. Sharma, A. K., Bhandari, R., & Pinca-Bretotean, C., (2020). A systematic overview on fabrication aspects and methods of aluminum metal matrix composites. In: *Materials Today: Proceedings* (Vol. 45, pp. 4133–4138). doi: 10.1016/j.matpr.2020.11.899.
31. Lee, C. S., Li, H., & Chandel, R. S. (1999). *Vacuum-free Diffusion Bonding of Aluminum Metal Matrix Composite.*
32. Zhang, X. P., Ye, L., Mai, Y. W., Quan, G. F., & Wei, W., (1999). Investigation on diffusion bonding characteristics of SiC particulate reinforced aluminum metal matrix composites (Al/SiCp-MMC). *Composites Part A: Applied Science and Manufacturing*, *30*(12), 1415–1421. doi: 10.1016/S1359-835X(99)00040-8.
33. Muratoğlu, M., Yilmaz, O., & Aksoy, M., (2006). Investigation on diffusion bonding characteristics of aluminum metal matrix composites (Al/SiCp) with pure aluminum for different heat treatments. *Journal of Materials Processing Technology*, *178*(1–3), 211–217. doi: 10.1016/j.jmatprotec.2006.03.168.
34. Hwang, Y. H., Horng, C. F., Lin, S. J., Liu, K. S., & Jahn, M. T., (1997). Effects of process parameters on properties of aluminum/stainless steel fiber metal matrix composites made by diffusion bonding. *Materials Science and Technology*, *13*(12), 982–988. doi: 10.1179/mst.1997.13.12.982.

CHAPTER 24

PERFORMANCE INVESTIGATION OF MIMO-OFDM SYSTEM IN NEXT GENERATION 5G COMMUNICATION: A REVIEW

BALRAM DAMODHAR TIMANDE, PANKAJ KUMAR MISHRA, SANDEEP D. BHAD, and PRADEEP KUMAR

Department of Electronics and Telecommunication Engineering, Rungta College of Engineering and Technology, Bhilai, Chhattisgarh, India

ABSTRACT

This chapter represents an overview of OFDM technology and its approach in the context of 5G wireless systems. The OFDM system can be used in both uplinking and downlinking in the case of the 5G system. The amalgamation of 'orthogonal frequency division multiplexing (OFDM)' and 'multiple-input multiple-output (MIMO)' antenna systems is gaining attention in the field of wireless communications. This combination promises improved 'quality of service (QoS)' and higher throughput to meet the needs of tomorrow. With this combination, better spectral efficiency and reliability of communication in the wireless environment can be achieved. In this chapter, we focus on the key issues that may be encountered in the MIMO-OFDM-based 5G wireless system. This chapter mainly focuses on the review analysis of various technologies used for PAPR reduction and error rate minimization in the OFDM system.

Computational Optimization, Modeling, and Simulation for Engineering Applications.
Anupam Shukla, Sourabh Rungta, Mohan Awasthy, and Rakesh Himte (Eds.)

24.1 INTRODUCTION

In the past decade, mobile and wireless communication have made noteworthy developments to interconnect the world with high-speed networks. Now the world is switching towards 5G and 6G with huge requirements of ultra-high speed and trustworthy wireless networks to fulfill the gap between the application requirement and technology. We need to introduce the next frontier technology which will stand in the future to serve the data security and speed requirements. Today's wireless networks should be capable of adopting all the possibilities and different technologies that may be faced in the near future, such as the application of software-defined radios, new error control schemes, new modulation schemes, and spectrum utilization. The limitations of the past generation motivate us to introduce the next-generation wireless network. The migration from the current generation to the next generation (5G) has many challenges since it deals with different wireless technologies, different platforms, and protocols. The objective of the next-generation system design is to make a powerful solution and to optimize the receiver complexity due to the integration of different technologies in the system. MIMO-OFDM systems are playing a key role in handling many issues in the current as well as future generations of wireless systems [1–3].

The multimode network integration is a basic practice which can access different network platforms such as smartphones, wireless gadgets, Wi-Fi modules, 3G and 4G LTE systems in parallel. While integrating the different platform technology, it should be taken care that the designed system is capable of handling user mobility. The 5G MIMO OFDM wireless system will make sure to reduce the system complexity with the integration of different platform networks [4].

This chapter gives the basic review about MIMO-OFDM based 5G wireless system. This chapter is planned under following heads, Section 24.2 gives the literature review analysis, it includes the papers based on MIMO-OFDM and 5G system. Section 24.3 presents about MIMO-OFDM system model, discussing about the MIMO antenna system with OFDM modulation. Section 24.4 gives about the PAPR reduction techniques. In Section 24.5, discussed about the expected results which met the proposed system. In last, Section 24.6 concludes about the chapter.

24.2 LITERATURE REVIEW

The following research articles have been reviewed and studied thoroughly. The proposed MIMO-OFDM system design for the 5G wireless system will be based on the problems identified during the review.

Ali et al. introduce the MIMO-OFDM system using RSTBC in a flat fading MIMO Channel. The author used 2×2 (Tx = 2 and Rx = 2 antenna) MIMO system and QPSK/16-QAM OFDM modulation technique for the improvement of BER. The author worked on the analysis of 2×2 MIMO-OFDM system performance with an experimental setup on an FPGA board. To cancel out interference between symbols in the OFDM system, a CP has been inserted in each symbol as a guard interval. It is observed from the simulation results that despite 10% to 20% loss, RSTBC is capable of recovering the original signal. Also, the BER performance with MMRC has been analyzed and it was found remarkable [1]. Ramteke et al. have introduced the limitations of the 3G and 4G wireless system and elaborated on the need for the next 5G wireless network. The authors have exposed the 5G features which can fulfill today's and tomorrow's system and customer requirements. The authors worked with MIMO with 802.11 based 5G wireless system [2]. Prabina et al. have proposed a collective client and antenna scheduling methodology exploiting spatial multiplexing with partial feedback for MIMO-OFDM downlink channel. The proposed algorithm overcomes the limitation of maximum CSI to the base station [5]. Mohammed et al. have worked on interference alignment (IA) in the MIMO channel. The authors considered two antenna selection norms such as MSR and MER. In order to evade subcarrier inequality across the multiple user antennas due to per sub-carrier selection, a controlled per sub-carrier antenna selection is employed. The authors found that spatial correlation can be minimized by modifying the remoteness between network nodes and the distance between the antennas inside each network node [6]. Abhijith et al. suggested MIMO cooperative scheme. The authors employed OFDM with MIMO in Cooperative scenarios. In this chapter, the group of subcarriers is selected and modulated using lower order modulation aided technique. The author found that the BER of OFDM-IM (Index Modulation) is improved using 8 QAM modulation techniques as compared to 16 QAM modulation [7]. Ngoc et al. proposed antenna selection techniques for the MIMO OFDM Wireless scheme. The authors employed 4x1 and 4x2 antenna systems to accomplish an improved energy efficiency performance. From the chapter review, it is noticed that based on the channel state a flexible antenna selection technique in which the number of active RF chains and the antenna indices are selected. The authors developed an efficient algorithm that results in optimal performance with lower complexity in the case of a small antenna system [8]. Hengyao Bao et al. proposed a PAPR reduction technique for the MIMO OFDM system.

The authors developed an adaptive PAPR reduction technique using the Bayesian approach by employing the surplus degrees-of-freedom of the transmit array. Results from simulation confirm a considerable performance improvement compared to existing/conventional methods in terms of PAPR reduction and complexity of computation [9]. Selahattin Gokceli et al. have proposed frequency selective PAPR reduction techniques. The authors explained a proficient PAPR curtailment technique using the iterative clipping and error filtering (ICEF) Algorithm. The proposed technique, where noise clipping has been efficiently done and effectively filtered within the transmitter passband, maintains the transmitted signal quality per PRB [10]. Shahid Manzoor et al. proposed the Channel estimation process. The authors analyze the response in different SNR conditions. It is shown that the adoption of the Reed Solomon FEC method improves the system performance compared to the Alamouti space time block code system [11]. S. S. Das et al. has proposed inter-carrier interference (ICI) minimization techniques in OFDM. This chapter focuses on maintaining the orthogonality so as to avoid the ICI. The fixed carrier bandwidth (FSB) has been adopted to minimize the ICI [12].

24.3 SYSTEM MODEL

24.3.1 MIMO SYSTEM MODEL

The MIMO antennas system is a fundamental technology in 4G/5G networks, offering a big advantage towards spectrum utilization. The spectrum efficiency can effectively be improved with multiple transmitting and receiving antennas. Deploying the additional antennas at the receiver or transmitter leads to an increase in signal power.

This could be a successful way to maximize the received power. The MIMO antenna system may have 'M' and 'N' number of antennas at the transmitter and receiver sections, respectively, as illustrated in Figure 24.1. Through the wireless channel, each antenna of the receiver accepts the direct as well as the indirect components that are anticipated for the neighboring antennas [13, 14]. It is shown by different authors that increasing the number of antennas at the transmitter and receiver gives a remarkable improvement in system performance in terms of throughput and BER. The spatial multiplexing due to an increase in the number of antennas in the MIMO system attains a linear growth in channel capacity.

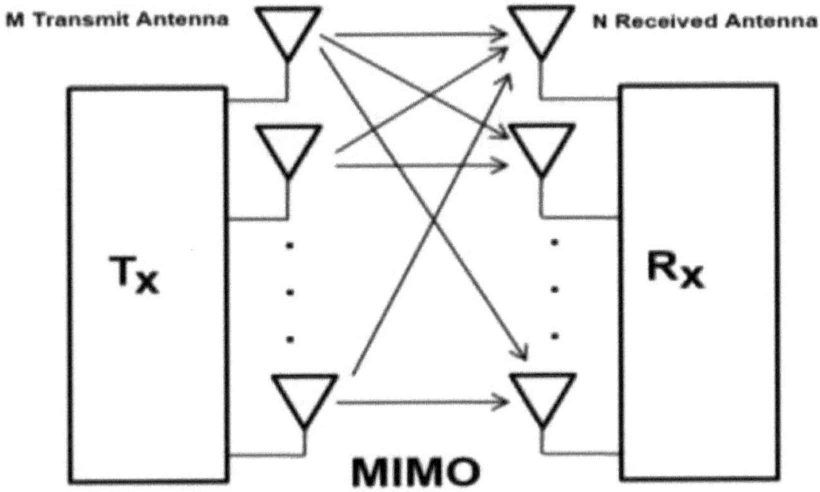

FIGURE 24.1 MIMO antenna system.

24.3.2 MIMO-OFDM SYSTEM

The model of MIMO-OFDM wireless communication shown in Figure 24.2 begins with the modulation of the message signal using BPSK, QPSK, QAM, etc. At this point, the outputs are serial complex numbers. In the next stage, a serial to parallel converter is used to provide parallel complex data. The IFFT or IDFT is used to convert the frequency domain signal into a time-domain signal.

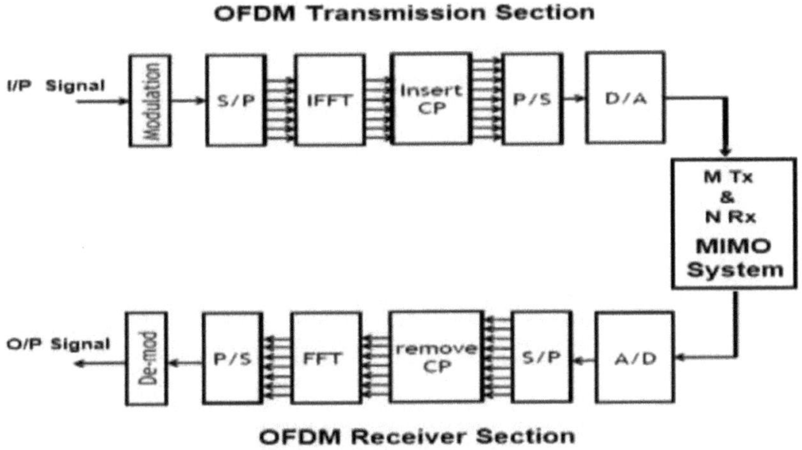

FIGURE 24.2 MIMO-OFDM system.

Using n-point IDFT or IFFT, the frequency separation (rate 1/ts) of the overall signal being transmitted over n subcarriers can be achieved [15]. Now the data transformed using IFFT are grouped into blocks as per the requirement of transmission subcarriers. The CP (Cyclic prefix) or guard signal is added in each block of data, called OFDM symbol. The objective of adding the CP is to overcome 'inter-symbol interference (ISI)' between OFDM symbols. After adding the CP in each block of data, the parallel output from IFFT is converted into the serial form using a parallel to serial converter. At this stage, the data is said to be OFDM modulated and is available in the time domain for transmission. Before transmission, DAC is used to convert the digital input into analog output in the time domain. The analog output of DAC is modulated with Radio Frequency for uplinking or transmitting through the wireless channels. For wireless transmission, the MIMO antenna systems are employed which consist of multiple transmitters (Tx) and receivers (Rx) antennas as shown in Figure 24.1. MIMO antenna system is used to provide antenna diversity, to enhance the system's reliability and the throughput or capacity. With/without transmitter antenna selection when the OFDM signal is transmitted through the MIMO antenna systems, then the signals go through the wireless channel and are received by receiving antennas of the MIMO system. After receiving the signal through MIMO receiving antennas, the received signal is converted into a digital domain using ADC. The output signal of ADC is converted back to parallel using a serial to parallel converter. At this stage, CP from each block is removed and the resulting signal is fed to the FFT block to convert the signal into the frequency domain. Before demodulation of the OFDM signal, the parallel time-domain signal is converted into serial form using a parallel to serial converter and then it is demodulated to obtain the original information [13, 14].

24.4 PAPR REDUCTION METHODOLOGY

OFDM has several properties which make it a trustworthy and robust technique used for high-speed transmission. The main disadvantage of OFDM is the presence of high PAPR in the transmitted signal. The ratio of the high peak signal power to the average signal power in the time domain is called peak to average power ratio (PAPR). Due to the logical summation of subcarriers signal at IFFT, high peaks are produced at the output. When such a signal with high peaks is given to the power amplifiers (PA) for amplification, it enters into saturation and behaves nonlinearly [16]. The PA starts amplifying the signal nonlinearly and thus resulting in high PAPR in the OFDM signal. The high PAPR results in interference between subcarriers and hence degrades the performance of the OFDM system. Similarly, due to high PAPR, the orthogonality of the OFDM signal fails, and that results in ICI. There are various schemes to improve the PAPR performance [17, 18], such as iterative companding transform and filtering (ICTF), ICEF algorithm, selective mapping (SLM), Bayesian approach, and many more.

24.5 EXPECTED RESULTS AND DISCUSSION

This chapter mainly focuses on the performance of the MIMO OFDM scheme on the system's error rates, channel capacity, spectral efficiency, and PAPR of the proposed system. In addition, the performance may include the reduction of interferences like ICI and ISI. The MIMO antenna system with Tx = 2, 3, 4, 5 and Rx = 2, 3, 4, 5 such combinations can be adopted to achieve high throughput, better spectral efficiency, and remarkable channel capacity. The channel capacity can be improved by increasing the number of antennas at the receiver and transmitter. The error rate can be measured with the implementation of antenna diversity schemes like MRC and Alamouti scheme. Also, the space-time code schemes like Alamouti STBC, STTC can be used to analyze error rates of the proposed system.

24.6 CONCLUSIONS

The blend of the MIMO antenna system with the OFDM scheme would be the best solution for upcoming wireless technology (5G), which would be the assured technology that fulfills the demand of the next-generation wireless networks. It is concluded from the review that antenna combinations like 2×4, 3×3, 4×5 would provide better throughput or capacity. The space-time coding scheme like Alamouti STBC would provide optimized error rates and the ICF scheme would be the better solution for lessening the PAPR.

24.7 FUTURE SCOPES

Future 5G technology may use different routing protocols. Power consumption will be a major issue in 5G networks, so optimizing power consumption will be a challenging task for future generations. The test of intelligence will be to choose the modulation scheme, the number of antennas, and the type of antennas to be used in 5G systems. End-to-end delay optimization will also be an important issue so researchers can employ different techniques to reduce end-to-end delay for fast and efficient communication.

KEYWORDS

- **5G networks**
- **antenna**
- **error rates**
- **generation**
- **interference**
- **MIMO system**
- **modulation**
- **OFDM**
- **wireless communication**
- **wireless technology**

REFERENCES

1. Alqahtani, A. H., Humadi, K., Sulyman, A. I., & Alsanie, A., (2019). Experimental evaluation of MIMO-OFDM system with rate less space-time block code. *International Journal of Antennas and Propagation, 2019.*
2. Ramteke, S. B., Deshmukh, A. Y., & Dekate, K. N., (2018). A review on design and analysis of 5G mobile communication MIMO system with OFDM. In: *2018 Second International Conference on Electronics, Communication and Aerospace Technology (ICECA)* (pp. 542–546). IEEE.
3. Kumar, P. S., Sumithra, M. G., Abilash, R., & Kumar, E. P., (2014). Performance analysis for differential STBC and SFBC MIMO-OFDM systems under different modulation schemes in a Rayleigh channel. In: *2014 International Conference on Electronics and Communication Systems (ICECS)* (pp. 1–5). IEEE.
4. Gupta, A., & Jha, R. K., (2015). A survey of 5G network: Architecture and emerging technologies. *IEEE Access, 3*, 1206–1232.
5. Pattanayak, P., Trivedi, V. K., Chakraborty, S., & Kumar, P., (2017). BER performance of multi user scheduling for MIMO-STBC and MIMO-OFDM broadcast network

with imperfect CSI. In: *2017 4th International Conference on Signal Processing and Integrated Networks (SPIN)* (pp. 66–70). IEEE.
6. El-Absi, M., Galih, S., Hoffmann, M., El-Hadidy, M., & Kaiser, T., (2015). Antenna selection for reliable MIMO-OFDM interference alignment systems: Measurement-based evaluation. *IEEE Transactions on Vehicular Technology, 65*(5), 2965–2977.
7. Nambi, S. A., & Giridhar, K., (2018). Lower order modulation aided BER reduction in OFDM with index modulation. *IEEE Communications Letters, 22*(8), 1596–1599.
8. Le, N. P., & Safaei, F., (2015). Antenna selection strategies for MIMO-OFDM wireless systems: An energy efficiency perspective. *IEEE Transactions on Vehicular Technology, 65*(4), 2048–2062.
9. Bao, H., Fang, J., Chen, Z., Li, H., & Li, S., (2016). An efficient Bayesian PAPR reduction method for OFDM-based massive MIMO systems. *IEEE Transactions on Wireless Communications, 15*(6), 4183–4195.
10. Gökceli, S., Levanen, T., Riihonen, T., Renfors, M., & Valkama, M., (2019). Frequency-selective PAPR reduction for OFDM. *IEEE Transactions on Vehicular Technology, 68*(6), 6167–6171.
11. Manzoor, S., Bamuhaisoon, A. S., & Alifa, A. N., (2015). Channel estimation for MIMO-OFDM systems. In: *2015 5th National Symposium on Information Technology: Towards New Smart World (NSITNSW)* (pp. 1–7). IEEE.
12. Das, S. S., De Carvalho, E., & Prasad, R., (2007). Variable sub-carrier bandwidths in OFDM systems. In: *2007 IEEE 65th Vehicular Technology Conference-VTC2007-Spring* (pp. 1866–1870). IEEE.
13. Paulraj, A. J., Gore, D. A., Nabar, R. U., & Bolcskei, H., (2004). An overview of MIMO communications-a key to gigabit wireless. *Proceedings of the IEEE, 92*(2), 198–218.
14. Goldsmith, A., Jafar, S. A., Jindal, N., & Vishwanath, S., (2003). Capacity limits of MIMO channels. *IEEE Journal on Selected Areas in Communications, 21*(5), 684–702.
15. Duman, T. M., & Ghraye, A., (2007). *Coding for MIMO Communication Systems.* John Wiley & Sons.
16. Cho, Y. S., Kim, J., Yang, W. Y., & Kang, C. G., (2010). *MIMO-OFDM Wireless Communications with MATLAB*. John Wiley & Sons.
17. Gupta, P., Singh, R. K., Thethi, H. P., Singh, B., & Nanda, S. K., (2019). Discrete cosine transform matrix based SLM algorithm for OFDM with diminished PAPR for M-PSK over different subcarriers. *Journal of Computer Networks and Communications, 2019.*
18. Amhaimar, L., Ahyoud, S., Elyaakoubi, A., Kaabal, A., Attari, K., & Asselman, A., (2018). PAPR reduction using fireworks search optimization algorithm in MIMO-OFDM systems. *Journal of Electrical and Computer Engineering, 2018.*

CHAPTER 25

A REVIEW OF BLOCKCHAIN AND ITS ENERGY CONSUMPTION

SNEHAL VAIRAGADE,[1] RAMESH KUMAR YADAV,[2] and DILIP MISHRA[2]

[1]*Department of Computer Science and Engineering, MATS University, Raipur, Chhattisgarh, India*

[2]*Faculty of Science and Technology, ICFAI University, Raipur, Chhattisgarh, India*

ABSTRACT

In the past decade, blockchain technology has surfaced as a means of addressing the privacy and complexity issues associated with distributed databases. By doing away with middlemen, it lowers costs for consumers and increases confidence in peer-to-peer communications. Due to its potential for business growth for corporations and its rapidly growing applications in collaborative smart-city ecosystems, healthcare, and governance, the notion of blockchain has undergone significant shifts over this time. Numerous platforms have been introduced, each with a unique design and consensus protocol. As a result, selecting the best platform for an application developer becomes difficult. Moreover, blockchain's objectives are at odds with those of a productive, environmentally friendly, cooperative digital ecosystem. As a result, closing this gap and creating new frameworks to integrate blockchain with those objectives become imperative. This chapter examines the development of blockchain technology, analyses energy use, and discusses the reasoning behind the evolution.

Computational Optimization, Modeling, and Simulation for Engineering Applications.
Anupam Shukla, Sourabh Rungta, Mohan Awasthy, and Rakesh Himte (Eds.)

25.1 INTRODUCTION

Professionals and everyone involved in socioeconomic life alike have given cryptocurrency a huge boost in popularity. In 2008, Nakamoto [1] published an article that aroused great interest from the scientific community. A year later, the author and a group of other researchers implemented the Bitcoin project (the first cryptocurrency). Unlike traditional currencies, where a third party (bank or financial institution) is required to enable a transaction to take place between two participants, cryptocurrencies use consensus models to validate transactions. The safety of the program and the privacy of the participants are realized through cryptographic methods.

To date, there are more than 4,000 cryptocurrencies with a market cap of about $2.48 trillion. This fact speaks of a new digital economy that is evolving every day. After Bitcoin's launch in 2009, cryptocurrency, with the rapid advancement of its technology involved, blockchain, has grown into a significant industry. Blockchain developments not only focus on improving its scalability, usability, and robustness, but also on improving users' privacy. Bitcoin, launched in early 2009, is the first decentralized virtual currency that removes the role of a central authority. A decade after Bitcoin started, the cryptocurrency-related businesses have become a multi-billion-dollar industry. The technology development has progressed so fast that cryptocurrency has now become a new virtual asset class. As we know, there are currently more than 4,000 cryptocurrencies available in the market and Bitcoin, the most valuable cryptocurrency, is traded at INR ₹2,735,235/– per coin (19/06/2021). Central authorities run traditional payment systems such as banks. However, cryptocurrency systems employ a different approach to remove the role of a central administration. Instead of keeping the transaction data private, cryptocurrency uses blockchain technology. In a blockchain system, each participant can verify all transactions in the system.

The most famous and widely used cryptocurrency in the world, Bitcoin, has been gaining popularity. It still retains the same fundamental design as when it was first founded in 2008, but as the global economy has changed repeatedly, there is now a far higher demand for cryptocurrencies than there was previously. Users are able to exchange value digitally without the intervention of a third party by using a cryptocurrency. Cryptocurrency operates on the premise that encryption methods can be cracked to produce a finite number of one-of-a-kind hashes. Users can trade hashes just like they would trade physical money thanks to a network of computers that verifies transactions.

Long-standing trade systems would need to be fundamentally modified to deal with this type of competition if cryptocurrencies became the standard for transactions on a global scale. Cryptocurrencies may be the single technology that has the greatest potential to upend the world's financial and economic systems because of this.

25.2 BLOCKCHAIN

Blockchain is a distributed ledger system for managing data and transactions that was initially created for the Bitcoin cryptocurrency. Since the concept of distributed ledger technology was first introduced in 2008, interest in it has risen. The central characteristics of Blockchain that provide security, anonymity, and integrity of information without any third-party company in charge of the transactions are the reason for the interest in it, and it therefore creates fascinating research fields, especially from the standpoint of technical problems and constraints.

Blockchain technology aims to establish a decentralized ecosystem in which data and transactions are controlled by no outside entity. Blockchain is a peer-to-peer database system that keeps an ever-growing list of data entries that are verified by the network nodes that are a part of it. Every activity ever made is detailed in the data, which is kept in a public ledger. Blockchain is a decentralized system that eliminates the need for a middle organization. Every node in the blockchain has access to and shares the details of every transaction that has ever been performed. Compared to centralized activities involving a third party, the system is more transparent due to this feature. Additionally, since all of the nodes in Blockchain are anonymous, it is more secure for other nodes.

Blockchain appears to be a good option for handling cryptocurrency transactions, but there are still certain technical issues and restrictions that need to be researched and resolved. To stop attacks and efforts to disrupt operations in Blockchain, high authenticity of transactions, security, and node privacy are required. Additionally, the Blockchain needs computer capacity to verify transactions.

25.2.1 CHARACTERISTICS AND FUNCTIONALITIES

1. **Decentralization:** The endpoints in a conventional centralized network must be verified by a reliable central server. This method raises the cost of computations and introduces a problem of

communication delay throughout the entire system. The blockchain is made up of peer-to-peer blocks that can be created without the assistance of a third party. It implies that numerous systems can be stored and updated on the blockchain without the need for a centralized server. All users or endpoints take part in transactions with such a decentralized service and slash the server in this distributed network.

2. **Persistency:** The blockchain verifies each transaction, and trustworthy miners preserve the transaction data. Winding back or deleting the transactions becomes challenging if they are on the list. These blocks cannot be altered because they are also verified by other miners.
3. **Auditability:** Users will be able to easily follow and verify the transaction information since every transaction in the blockchain will have a digital signature from the sender that saves the block with the date.
4. **Anonymity:** Data security in the blockchain is achieved through the use of asymmetric encryption methods. All transactions are approved to verify the recipient. To enable a self-generated email, the communicator connects with the blockchain and creates a distinct set of addresses, keeping their identity a secret. As a result, a centralized ownership will maintain the users' actual identity and minimize the revelation of the sender's identity.
5. **Autonomous:** We can publish the signed nodes and review them if other nodes in the decentralized system accept them because no single organization controls the blockchain-based system. In order to ensure that transfer of data will take place securely in the blockchain, the consensus base was created by each node in the network approving a node.
6. **Immutability:** Before being added to the block, the transaction data is verified. The transactions are forever recorded on the blockchain. The block's data cannot be altered. Since the data in each block is connected by the hash key and changes to the data invalidate subsequent blocks, any attempt to change the data would be quickly detected.
7. **Transparency:** Blockchain is a distributed ledger system that allows all users to publish their records and access the data from the nodes. With this blockchain technology, the system logs and keeps track

of transaction records in an accessible distributed ledger. All nodes present in the same network can access this information because it is open and trustworthy.

8. **Traceability:** Hashing methods are used to encrypt the data in blockchain blocks. There will be hash keys for each block. Each block in the network is connected to the preceding blocks by their hash keys. As a result, the blockchain network makes it easy to trace a block using its hash key.

25.2.2 TYPES OF BLOCKCHAIN

Based on the availability and accessibility to users, blockchain is divided into three main forms:

1. **Public (Permission Less) Blockchain:** Any node can join the network and participate in the processing, storing, and validating of the transaction records through a consensus algorithm on a public blockchain, which is a decentralized, open ledger.
2. **Private (Permissioned) Blockchain:** The blockchain network is a constrained network where no one can easily join. It is a specific kind of centralized blockchain that is accessible under the management of a single entity. The public is only given limited access to the private blockchain's data read authorization. Limited organizations or niche markets are the only ones eligible for private blockchain. Different use cases for the private blockchain include assets, logistics management, digital identity, and vote counting.
3. **Consortium Blockchain:** A chain that is only partially decentralized is the collaborative blockchain. The well before node will be given the power to decide in advance on the kind of service. The blockchain transactions may be accessible to the remaining nodes, but the consensus mechanism is not.

25.2.3 PHASES OR GENERATIONS OF BLOCKCHAIN

Three generations make up the development of blockchain technology. Blockchain 1.0 is the first generation of cryptocurrency, Blockchain 2.0 is the second generation and is known as the modern economy, and Blockchain 3.0 is the third generation and is known as the digital age.

25.2.4 WORKING OF BLOCKCHAIN

Building blocks holding data are suggested by distributed ledger technology, which is connected to the chain. Through each block holding the previous block header, blocks are connected to one another. The hash key will change if some information in the preceding block is altered, resulting in a mismatch in the hash key throughout the chain of blocks. It stops data from being altered. A block will be used to describe a process when a user wants to transfer some transaction data to someone else. To add the blockchain block, the block must be broadcast to every other node of the network. The transaction needs to be approved by the node's miners. When a block is formed, the miners are given the right to validate it by solving numerically difficult problems. After verification, the block is put onto the network, concluding the transaction. Choosing the user who posted the following block is the next step. A network is created by joining the chain of blocks that have been verified.

25.3 LITERATURE REVIEW

The environment is seriously concerned about the amount of energy used by the blockchain network's computationally intensive consensus procedures. As it is the most thoroughly verified protocol and guarantees a high degree of security, the majority of the blockchain system uses the computationally demanding PoW protocol. The Bitcoin community will require the same amount of energy that the rest of the world does now by 2020, according to a report by the American magazine Grist. This energy use puts the planet in danger by leaving carbon footprints and contributing to global warming [13]. To solve this problem, a number of academics suggested using an alternative consensus mechanism or energy-efficient equipment. The voting- or capability-based protocols take the role of the computationally intensive protocols. A new generation equipment accelerator that Intel proposed, and patents attempts to cut the Bitcoin network's mining power by 35% [19]. Numerous initiatives are being taken to lower the energy usage of the distributed ledger technology while maintaining its benefits, like decentralization, privacy, and safety. Studying the usage of sustainable energy for mining, though, would be intriguing. Throughput before being processed, every transfer in a blockchain system needs to pass peer-to-peer verifications. With more users, this takes longer, particularly in a public blockchain system where each user must validate each transaction. As a result, a blockchain network can process far more transactions per second than current centralized systems. Researchers and innovators have been focusing on improving the blockchain technology's performance when utilized extensively. Contemporary consensus techniques, including the delegated proof of stake [10], Byzantine fault tolerance [11], and the federated Byzantine agreement [12] were proposed and used by different blockchain platforms [14, 19]. To address the issue of activities not being processed in a chain until the block is finished being mined by the miners, verified by the verifiers, and modified in the blockchain, it was suggested that linked activities be used instead of linked blocks in the ledger. The record is only shared among the peers taking part in an operation in the linked-transactions approach. As a result, various ledgers exist in the same network. As a result, it is challenging to determine the right ledger state when one peer's ledger is compromised or the peer acts maliciously. This has an effect that is comparable to the specific point of failure problem in a centralized participant network. The blockchain development platform addresses the single point of failure in a linked-transactions ledger. Brown et al. [15, 16] presented a node-maintained ledger with transactions linked in a directional acyclic graph. In IOTA, the directed graph of exchanges is referred to as a "tangle" with the graph's vertices denoting transactions and its edges denoting approvals (also known as approvals). A tip is a transaction that hasn't received any node approval. A user must choose two tips for the acceptance of the deal submission in order to be accepted into the network [17]. A trunks activity and a branch activity are the names given to these tips. Moreover, given that it is feasible for a potential hacker to generate randomized transactions to choose a tip and execute invalid interactions, tangled transfers are more vulnerable to active attacks operations.

25.4 CONCLUSION

Each exchange in a distributed ledger like blockchain must undergo peer-to-peer confirmations before it can be completed. This requires longer when there are more users, particularly on public blockchain platforms where every user has to validate every transaction. Consequently, compared to existing centralized systems, a distributed ledger network may execute many more operations per second. The efficiency of blockchain technology under heavy use has been a focus for researchers and developers. New consensus techniques like the delegated proof of stake and the Byzantine fault tolerance [10], and the federated Byzantine agreement [12] were proposed and used by different blockchain platforms. Schwartz, Youngs, & Britto [14, 18] proposed to solve the problem of the transactions which are being delayed for processing in a block until the block is complete by using linked transactions instead of linked blocks in the ledger, mined by the miners, verified by the validators, and updated in the ledger. However, in a transaction, the ledger is shared only among the peers who are participating under the linked-transactions solution, which causes an accumulation of many ledgers in the same network. Incidentally, if any ledger gets corrupted due to any reason, it is very difficult to agree on the correct state of the ledger, resulting in a similar issue to the single point of failure issue in a concurrent peer-to-peer network. To overcome the issue arising due to a single point of failure in a linked-transactions ledger, a ledger of transactions connected in a directed acyclic graph was proposed by the blockchain development platform IOTA [15, 16] and maintained at each node. The directed graph of transactions is called a tangle in IOTA, where the vertices of the graph represent the transactions and the edges represent the validation. An unapproved transaction (Tip) can be approved in the network if a user selects two tips (trunk transaction and a branch transaction) for the approval of the transaction submission [17, 18]. However, the tangled transactions are more prone to Sybil attack because it may lead to creating random transactions by any suspicious attacker to select a tip, and it can then result in invalid transactions.

KEYWORDS

- **blockchain technology**
- **consensus**
- **hashing algorithms**
- **replication**
- **scalability**
- **security**

REFERENCES

1. Bahl, L. R., Brown, P. F., De Souza, P. V., & Mercer, R. L., (1989). A tree based mostly applied math language model for language speech recognition. In: *Acoustics, Speech and Signal Process, IEEE Transactions* (Vol. 37, No. 7, pp. 1001–1008). Yorktown Heights, NY, USA.

2. Swan, M., (2015). *Blockchain: Blueprint for a New Economy*. O'Reilly Media, Inc.
3. Pearl Brereton, Barbara A. Kitchenham, David Budgen, Mark Turner, & Mohamed Khalil (2007). Lessons from applying the systematic literature review process within the software engineering domain. *Journal of Systems and Software, 80*(4), April 2007, 571–583.
4. Coinmarketcap, (2016). *Crypto-Currency Market Capitalizations*.
5. https://coinmarketcap.com/ (accessed on 21 December 2023).
6. Nakamoto, S., (2008). *Bitcoin: A Peer-to-Peer Electronic Cash System, Consulted, 1*(2012), 28.
7. https://blockchain.info/charts/transactions-per-second (accessed on 21 December 2023).
8. https://blockchain.info/charts/avg-confirmation-time (accessed on 21 December 2023).
9. https://github.com/ethereum/wiki/wiki/Sharding-FAQ (accessed on 21 December 2023).
10. Aste, T., Tasca, P., & Matteo, T. D., (2017). Blockchain technologies: The foreseeable impact on society and industry. *Computer, 50*(9), 18–28.
11. Salah, K., Rehman, M. H. U., Nizamuddin, N., & Al-Fuqaha, A., (2019). Blockchain for AI: Review and open research challenges. *IEEE Access, 7*, 10127–10149.
12. Larimer, D., (2014). *Delegated Proof-of-Stake Consensus*. Available online: https://bitshares.org/technology/delegated-proof-of-stake-consensus (accessed on 21 December 2023).
13. Castro, M., & Liskov, B., (1999). Practical byzantine fault tolerance. In: *Proceedings of the 3rd Symposium on Operating System Design and Implementation (OSDI)* (pp. 173–186). New Orleans, LA, USA, Unisex Association: Berkeley, CA, USA.
14. Schwartz, D., Youngs, N., & Britto, A., (2014). *The Ripple Protocol Consensus Algorithm* (pp. 5, 8). Ripple Labs Inc White Pap.
15. *Bitcoin Could Cost Us Our Clean-Energy Future*. Grist. Available online: https://grist.org/article/bitcoin-could-cost-us-our-clean-energy-future/ (accessed on 21 December 2023).
16. Brown, R. G., Carlyle, J., Grigg, I., & Hearn, M., (2016). *Corda: An Introduction*. R3 CEV 2016, 1, 15. 198. Corda. Available online: https://docs.corda.net/ (accessed on 21 December 2023).
17. Popov, S., (2016). *The Tangle*, p. 131. Available online: https://www.iota.org/foundation/research-papers (accessed on 21 December 2023).
18. *What Is IOTA*. Available online: https://docs.iota.org/introduction/what-is-iota (accessed on 21 December 2023).
19. *Tip Selection—IOTA Docs*. Available online: https://docs.iota.org/introduction/tangle/tip-selection (accessed on 21 December 2023).

CHAPTER 26

INVESTIGATION OF INVERSION IN PLANAR KINEMATIC CHAINS

KUNAL DEWANGAN[1] and RAHUL CHOWDHURY[2]

[1]*Mechanical Engineering, Kalinga University, Naya Raipur, Chhattisgarh, India*

[2]*Mechanical Engineering Department, Rungta College of Engineering and Technology, Bhilai, Chhattisgarh, India*

ABSTRACT

This study addresses the challenge of locating inversions within planar kinematic chains, specifically focusing on six-link one-degree-of-freedom, eight-link one-degree-of-freedom, nine-link two-degree-of-freedom, and ten-link three-degree-of-freedom kinematic chains. The findings of this research offer valuable insights for designers, aiding in the prevention of kinematic chain duplication during the conceptual phase of design. In this methodology, a path matrix (PM) is constructed using parameter-based values. Inversions are identified by examining the kinematic chain string (KCS) derived from the path matrix (PM). This approach enhances the understanding of kinematic chain structures and provides a foundation for design optimization in engineering applications.

26.1 INTRODUCTION

The process of creating new kinematic chains and systems is known as kinematic synthesis. All prospective kinematic chains are categorized according to their number of links, degrees of freedom, and joint types as part of the

Computational Optimization, Modeling, and Simulation for Engineering Applications.
Anupam Shukla, Sourabh Rungta, Mohan Awasthy, and Rakesh Himte (Eds.)

structural synthesis of kinematic chains. It could be as easy to comprehend mechanisms' roles as to comprehend their kinematic structure. In kinematic chains and processes, an understanding of the structure will lead to conscious list improvement, evidence discrimination, order, and inversion detection. It also results in the identification of the ideal chain for acquiring a specific collection of structural traits.

Based on Davies and Crossley [1] and Franke's [2] they developed the fundamental union technique, which led to the batch synthesis of 40, 9-link 2-DOF kinematic chains and 230, 10-link 3-DOF kinematic chains.

Using Franke's notation, Hass & Crossley [3] created a class of four degrees of freedom kinematic chains. Franke's language was also used by Soni [4] to develop two broad constraint kinematic chains. In order to compare kinematic chains and inversion for stiffness, compactness, appropriateness, and mechanical advantage for application as platform type robots, Rao [5] provided criteria and measurements.

Peter Mitrouchev [6] is in charge of the morphological bound determination for kinematic instruments in mechanical technology. It is decided by sub-chain structural balances that take into consideration the edge and end-effector positions of the robot. There are established collections of instruments that are consistently balanced. Destroying the balanced ones thereby limits the range of potential designs.

According to Peter Mitrouchev [7], he has made significant contributions to the synthesis of unique degrees of freedom for planar and spatial mechanisms. MMT kinematic graphs are used to provide a significant structure in some contemporary industrial robots with planar chains. To calculate the number of actuator attachment options for planar pin-jointed primary motivating mechanisms in robotics, Mitrouchev is interested in a formulation. It is founded on structural sub-chain balances that account for the location of the actuator on the robot. Groups of groupings with typically even components can be identified. In order to be considered for use as robot controllers, closed kinematic chains with degrees of freedom (d.o.f.) greater than one are believed to benefit most from parallelism, according to Rao [8].

There are many alternative chains for a given number of connections and d.o.f., thus the designer must have evidence to select the chain that has the best chance of achieving the desired result. As demonstrated by Rao & Srinath [9], the structural parallelism between both the links of a kinematic chain significantly affects their relative displacements. Greater mechanical advantage and more accurate function generation may be achieved by carefully choosing the input and output links. The best possible chain can be chosen using this method from a large pool of distinctive chains that all have the same number of links and degree of freedom.

A joint-based technique for isomorphism detection in planar kinematics was presented by Bedi & Sanyal [10]. Planar kinematic chains' topological characteristics are identified, as well as those of their inversions. The suggested joint-based method consists of three steps: determining the connection of joints at all levels, determining isomorphic chains based on the connectivity of joints at all levels, and determining distinct inversions. Planar kinematic chains' topological characteristics can be determined using the modified joint connectivity approach first proposed by Bedi & Sanyal [11].

To detect isomorphism and inversions, Dewangan & Shukla [14], a least distance matrix (LDM) is constructed using the parameter values. Isomorphism is found using the chain string (CS) of the LDM. Grounding a different link in the kinematic chain causes an inversion.

26.2 INVERSION IN PLANAR KINEMATIC CHAINS

When different links in a kinematic chain are fixed one at a time, numerous mechanisms are built through the process of inversion.

Following are the terminologies are used [13]:

1. **Distance Matrix:** The sum of the distances between each link in a chain yields a distance matrix D. All of the diagonal elements in the matrix will have values of 0, and Dii will have the least number of joints between links I and k.
2. **Node Number:** A node of a link is given a node number, which is a numeric value. The ratio of nodes to the number parameter at a particular link is the node number (Figure 26.1).

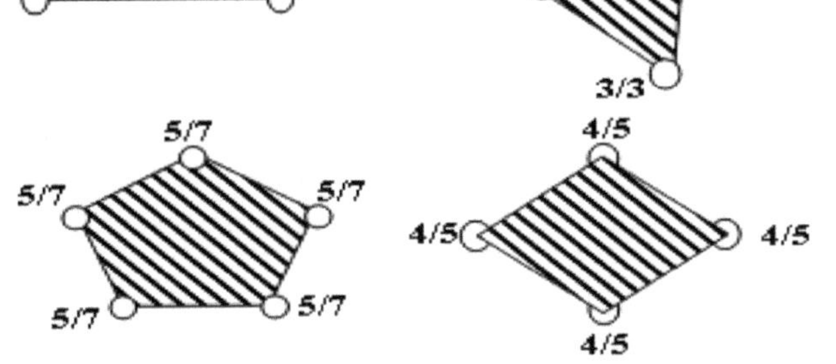

FIGURE 26.1 Node number for links [13].

3. **Joint Code:** It is defined as a multiple of the number of nodes that link at a specific joint (Figure 26.2).

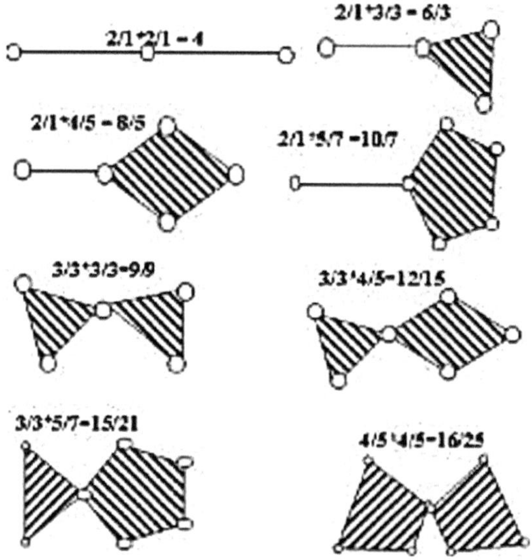

FIGURE 26.2 Joint codes of every potential link combination [13].

4. **Kinematic Link Value (KLV):** The distance matrix's entries in each row add up to a particular value. This is known as the KLV for that specific link.
5. **Kinematic Link String (KLS):** In ascending order, it refers to the components of a particular row (or column) of the distance matrix.
6. **Kinematic Chain String (KCS):** This formula for a specific chain is calculated by adding all of its KLVs as well as all of its matrix elements.
7. **Total Chain Value String (TCVS):** This formula for a specific link is obtained by adding the elements of each row value of that connection.

26.2.1 SIX LINK ONE DEGREE OF FREEDOM

The strategy includes a parametric methodology for identification of isomorphism. The strategy is clarified utilizing Watt and Stephenson's chain. A separation framework (DM) is shaped utilizing the qualities based on the parameter. Watt chain is appeared in Figure 26.3. Connection An and D are ternary connection and connection B, C, D, and E are twofold connection. A hub code of "2" is assigned to each hub of a double connection, and a joint code of "3/3" is assigned to hubs of ternary connections. The node codes for quaternary links and pentaternary links, respectively, were 4/5 and 5/7.

For the sake of simplicity in calculating a non-fragmentary worth, an incentive based on taking L.C.M. of joint characteristics is assigned to each joint, allowing us to use the same denominator for all joint codes for varying connection availability.

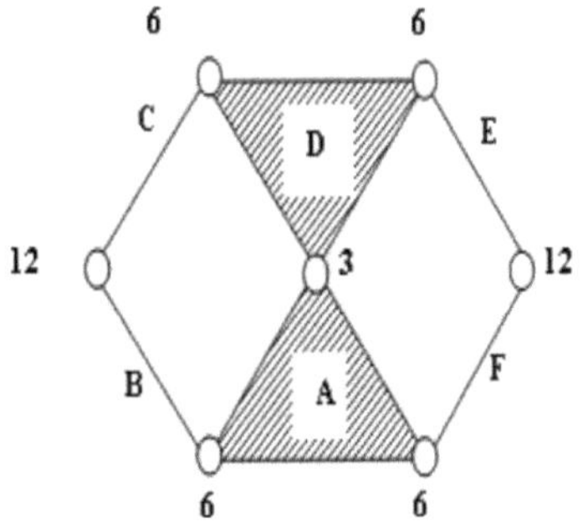

FIGURE 26.3 Watt chain [13].

A node that combines two binary links, one binary and one ternary links, and two ternary links will be given the node number 12/3, 6/3, 3/3, as seen in Figure 26.3.

The distance matrix (DM) for the Watt chain is shown below:

	A	**B**	**C**	**D**	**E**	**F**	
A	0	6	9	3	9	6	**33**
B	6	0	12	9	15	12	**54**
C	9	12	0	6	12	15	**54**
D	3	9	6	0	6	9	**33**
E	9	15	12	6	0	12	**54**
F	6	12	15	9	12	0	**54**
							324

Ternary link A is linked to binary link B, as shown in the matrix for the Watt chain, and the node number is "6." The shortest distance connecting links A and C is evaluated for connection, i.e., 6+3=9. Similarly, the joint code representing connectivity between links A and D will be "3," and so on. For a six-link watt chain, a 6x6 matrix will be created.

Kinematic link string (KLS) values for links A, B, C, D, E, and Fare are as follows: [0,2(6),3,2(9)], [0,2(12),6,9,15], [0,6, 9,15,2(12)], [0, 2(6), 3, 2(9)], [0, 2(12),6,9,15], and [0, 2(12),6,9,15].

As shown in Figure 26.3, the KLV for links A, B, C, D, E, and F are 33, 54, 33, 54, and 54, respectively, and the kinematic chain string (KCS) for Watt chain is 324, 2 [1(3), 4(6), 4(9), 4(12), 2(15)]. Distance matrix (DM) is shown below:

	A	**B**	**C**	**D**	**E**	**F**	
A	0	6	12	18	6	6	**48**
B	6	0	6	12	12	12	**48**
C	12	6	0	6	18	6	**48**
D	18	12	6	0	12	12	**60**
E	6	12	18	12	0	12	**60**
F	6	12	6	12	12	0	**48**
							312

As illustrated in the matrix, ternary link A is linked to binary link B for the Stephenson chain, and the node value is "6." For connectivity between links A and C, the smallest distance is evaluated, 6+6=12, and for connectivity between connect An and interface D, the shortest separation is considered, 6+12=18. A 6x6 lattice will be formed for a 6 connection Stephenson chain.

[0,3(6),12,18], [0,2(6),3(12)], [0,3(6),12,18], [0,6,3(12),18], [0,6,3(12), 18], [0,6,3(12),18], and [0,2(6),3(12)] individually. The KLV for connect A, B, C, D, E, and F in the Stephenson chain is – 48, 48, 48, 60, 60, and 48 separately, as shown in Figure 26.4. The KCS is 312, 2[6(6), 7(12), 2(18)] when all the KLVs of a Stephenson chain are added together. Watt chain have two distinct inversions and Stephenson chain have three distinct inversions.

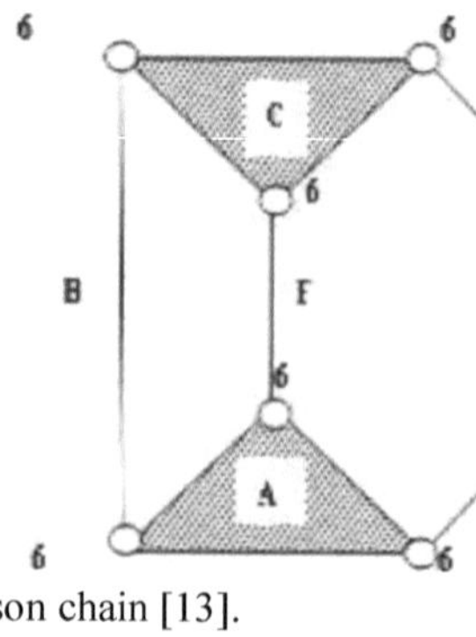

FIGURE 26.4 Stephenson chain [13].

26.3 RESULTS

The technique is used to kinematic chains with 10 links of one and three degrees of freedom, nine links of two degrees of freedom, and eight links of one degree of freedom.

Table 26.1 shows the results of inversions for an eight-link one degree of freedom kinematic chain (Appendix I), and from Table 26.1, after comparing each KCS and KLV for 16 kinematic chains, 71 inversions were obtained.

TABLE 26.1 Kinematic Chain Inversions with Eight Links of One Degree of Freedom

KCN	KLV	KLS	Distinct Inversion
1	1200	2(75),2(150),2(225),300	2
	2025	150,2(225),2(300),375,450	
2	1200	2(75),2(150),2(225),300	2
	2150	150,2(225),2(300),2(475)	
3	1800	2(150),2(225),300,2(375)	4
	1350	75,3(150),225,2(300)	
	1125	2(75),2(150),3(225)	
	1950	150,2(225),2(300),2(375)	
4	1725	2(150),2(225),2(300),375	8
	1350	75,3(150),2(225),375	
	1200	2(75),2(150),2(225),300	
	2100	150,2(225),300,2(375),450	
	2025	150,225,3(300),2(375)	
	1425	75,3(150),2(225),450	
	1125	2(75),2(150),3(225)	
	1800	2(150),2(225),300,2(375)	
5	1350	75,3(150),2(225),375	4
	2025	150,3(300),225,2(375)	
	1800	2(150),2(225),300,2(375)	
	1200	2(75),2(150),2(225),300	
6	1875	2(150),225,3(300),450	5
	1425	75,4(150),2(300)	
	2025	150,225,4(300),450	
	1275	75,4(150),2(300)	
	975	3(75),4(225)	

TABLE 26.1 *(Continued)*

KCN	KLV	KLS	Distinct Inversion
7	1500	75,2(150),2(225),300,375	2
	1800	2(150),2(225),300,2(375)	
8	1725	2(150),2(300),2(225),375	2
	1425	75,2(150),2(225),2(300)	
9	450	25,3(50),75,2(100)	6
	575	2(50),75,4(100)	
	600	3(50),3(100),150	
	650	2(50),75,2(100),125,150	
	425	2(25),50,3(75),100	
	575	2(50),2(75),2(100),125	
10	330	2(20),2(40),3(70)	5
	560	40,2(60),80,100,2(110)	
	660	50,70,90,2(100),110,140	
	370	20,40,2(50),2(60),90	
	590	2(50),70,2(100),2(110)	
11	440	20,3(40),70,90,140	8
	600	40,60,2(80),100,110,130	
	800	50,3(100),150,160	
	580	3(50),90,100,110,130	
	400	20,2(20),3(60),100	
	690	50,70,2(100),2(110),150	
	630	40,60,2(80),100,110160	
	520	40,50,60,2(80),100,110	
12	1110	60,3(120),2(210),270	7
	1560	120,150,180,2(240),300,330	
	1860	3(150),270,300,390,450	
	1710	2(150),210,270,2(300),330	
	1200	60,2(150),3(180),300	
	2070	150,210,2(300),2(330),450	
	1800	120,180,2(240),300,330,390	

TABLE 26.1 *(Continued)*

KCN	KLV	KLS	Distinct Inversion
13	1440	4(120),2(270),420	7
	1920	120,3(240),300,390	
	2145	150,225,2(300),2(375),420	
	1485	75,2(150),2(225)270,390	
	1475	75,2(150),2(225),270,390	
	1590	120,150,225,3(240),375	
	1515	120,150,225,3(240),300	
14	1440	4(120),2(270),420	4
	1680	120,150,3(240),300,390	
	1800	3(150),270,300,2(390)	
	1920	2(150),4(300),420	
15	1032	48,3(120),2(168),288	2
	1632	120,168,2(240),2(268)	
16	1440	4(120),240,2(360)	3
	1980	120,3(240),300,360,480	
	1440	2(120),5(240)	
Total Inversions			*71*

26.4 SUMMARY AND CONCLUSION

1. In this chapter, a straightforward, efficient, and trustworthy method for inversion detection is suggested.
2. With this approach, mechanical kinematic chain inversions are easily found.
3. The structure of a fundamental jointed planar kinematic chain is inverted as best as it can be here using a broad methodology.
4. This method is simple to use as the number of connections and degrees of freedom increase, and with a little tweaking, it can be adjusted to work for 'n' links and 'f' degrees of freedom.
5. To save time and get results more quickly, such a procedure can also be automated and coded.

KEYWORDS

- **degrees of freedom**
- **inversion detection**
- **kinematic chains**
- **kinematic link value**
- **matrix**
- **planar kinematic chains**
- **Stephenson chain**

REFERENCES

1. Davies, T. H., & Crossley, F. R. E., (1966). Structural analysis of planar linkages by Franke's condensed notation. *Journal of Mechanisms, 1*(2), 171–183.
2. Franke, R. (1958). Vom Aufbau der Getriebe. VDI-Verlag. (Vol. 1, 2). Beuthvertrieb, Berlin, VDI, Düsseldorf.
3. Haas, S. L., & Crossley, F. R. E., (1969). Structural synthesis of a four-bit binary adding mechanism. *J. Eng. Indust., ASME Trans, Series B, 91*, 240–250.
4. Soni, A. H., (1971). Structural analysis of two general constraint kinematic chains and their practical application. *ASME Journal of Engineering for Industry, 93B*(1), 231–238.
5. Rao, A. C., (1995). Topological characteristics of linkage mechanisms with particular reference to platform type robots. *Mechanism and Machine Theory, 30*(1), 33–42.
6. Peter, M., (2006). Sub-chain symmetry approach for morphological choice of planar mechanisms in robotics. *European Journal of Mechanics A/Solids, 25*, 166–188.
7. Peter, M., (2008). Formulation for actuators' number enumeration for main planar structures in robotics. *European Journal of Mechanics A/Solids, 27*, 622–646.
8. Rao, A. C., (2004). Parallelism in planar kinematic chains (manipulators). *Mechanism and Machine Theory, 39*, 1111–1122.
9. Rao, A. C., & Srinath, A., (2007). Planar linkages: Structural influence on mechanical advantage and function generation. *Mechanism and Machine Theory, 42*, 472–481.
10. Sanyal, S., & Bedi, G. S., (2010). Joint connectivity: A new approach to detection of isomorphism and inversions of planar kinematic chains. *J. Inst. Eng. (India), 90*, 23–26.
11. Sanyal, S., & Bedi, G. S., (2011). Modified joint connectivity approach for identification of topological characteristics of planar kinematic chains. *Journal of Mechanical EngineeringScience, 225*(11), 2700–2717.
12. Mruthyunjaya, T. S., & Balasubramanian, H. R., (1987). In quest of a reliable and efficient computational test for detection of isomorphism in kinematic chains. *Mechanism and Machine Theory, 22*(2), 131–139.
13. Kunal, D., et al., (2021). Detection of isomorphism in planar kinematic chains. *International Journal of Scientific Research and Engineering Development, 4*(4), 490–495.
14. Dewangan, K., & Shukla, A. K., (2022). A parametric approach to detect isomorphism and inversion in the planar kinematic chains. In: *Machines, Mechanism and Robotics* (pp. 743–818). Springer, Singapore.

APPENDIX I

Eight Link One Degree of Freedom Kinematic Chain

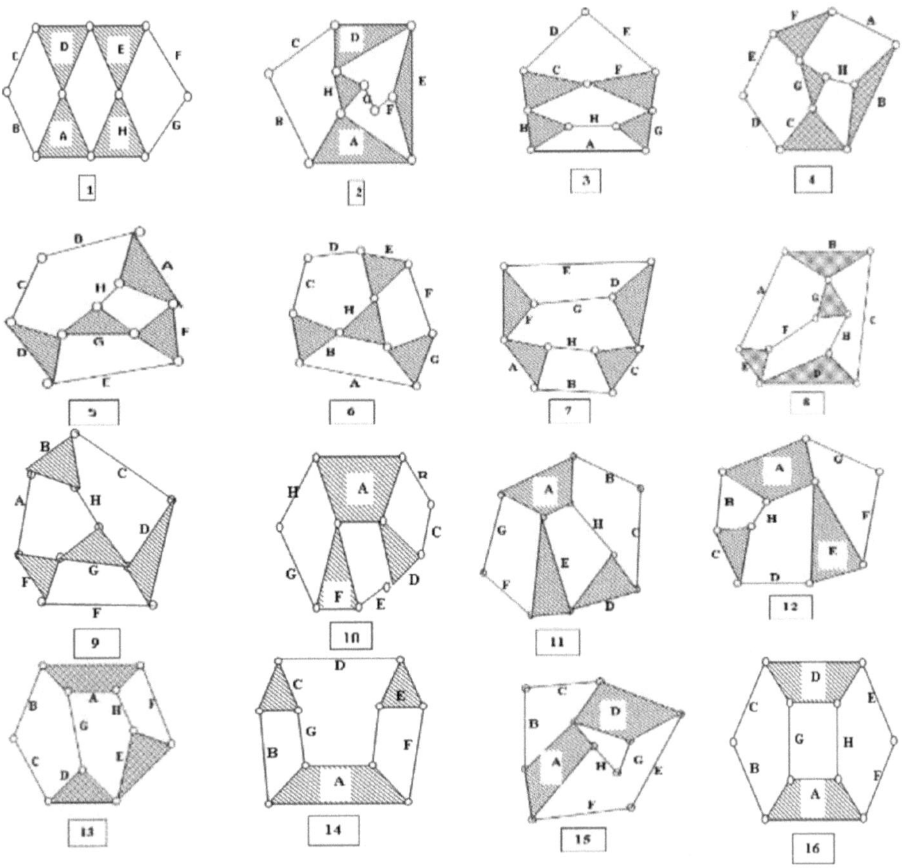

CHAPTER 27

LIQUID STATE PROCESSES FOR FABRICATION OF ALUMINUM METAL MATRIX COMPOSITES FOR AUTOMOTIVE AND AIRCRAFT APPLICATIONS

RITYUJ SINGH PARIHAR,[1] AGNIVESH KUMAR SINHA,[2] KASI RAJA RAO,[2] GULAB PAMNANI,[3] and VINAY SONI[4]

[1]*Mechanical Engineering, Shri Shankaracharya Institute of Professional Management and Technology, Raipur, Chhattisgarh, India*

[2]*Mechanical Engineering Department, Rungta College of Engineering and Technology, Bhilai, Chhattisgarh, India*

[3]*Mechanical Engineering Department, Malaviya National Institute of Technology Jaipur (MNIT), Rajasthan, India*

[4]*Mechanical Engineering Department, Columbia Institute of Engineering and Technology, Raipur, Chhattisgarh, India*

ABSTRACT

In large part due to the excellent strength to weight ratio that they possess, Aluminum Metal Matrix Composites, also known as Al-MMCs, have gained popularity for their uses in the automotive and aerospace industries. This makes it easier to achieve the goal of achieving a reduced weight for aviation or automotive components, which in turn minimizes the amount of fuel consumed and increases performance. The various liquid-state processing

Computational Optimization, Modeling, and Simulation for Engineering Applications.
Anupam Shukla, Sourabh Rungta, Mohan Awasthy, and Rakesh Himte (Eds.)

processes of aluminum metal matrix composites (Al-MMCs) for utilization in aviation and automobiles are discussed in this chapter. In addition to this, this sheds light on the significant research gaps that were discovered throughout the literature assessment.

27.1 INTRODUCTION

Composite systems, specifically metal matrix composites (MMCs), have been researched and investigated since the 1970s. However, the research and optimization of MMCs has taken longer to establish compared to metals and polymers. MMCs are modern materials that combine the toughness of a metal matrix, such as aluminum, with the hardness of ceramic reinforcement to create composites with exceptional material properties. MMCs have been studied for many years, and their potential advantages over conventional alloys are increasingly recognized. Recent trends show that MMCs are favored and viable for a wide range of applications in modern engineering. However, there are times when MMCs can be expensive to develop and do not offer the necessary repeatability for manufacturing. Over time, advancements in manufacturing, testing, and characterization technologies have improved our understanding of MMCs. These advancements have made it possible to fabricate composite systems using carbon nanotubes, ceramics, and graphite matrix composites [1, 2].

Nowadays, research and development of MMCs have reached a level where optimized fabrication techniques and abundance of sustainable knowledge make these advanced materials a feasible preference for lower weight applications instead of traditional choices. Although MMCs have been examined and analyzed for a long time, the adoption of these materials is still very limited due to the cost factor. The cost is the main hurdle in the rapid adoption in engineering applications due to the lack of better understanding as well as manufacturing techniques. Additionally, the higher cost also applies to the production of fiber and particulate reinforcement. The foreseeable driving force for lightweight and highly efficient transport methods has enhanced exploration in composite materials, facilitating the development of low-cost materials and manufacturing techniques. The recent manufacturing techniques include automatic equipment which allows for more reliable, cost-effective, and repeatable manufacturing that did not exist more than a few years ago. Several mixtures of matrix as well as reinforcement are fabricated and characterized for MMCs. The numerous matrix materials consist of aluminum, titanium, magnesium, zinc, and copper have been examined and a compilation of reinforcements such as alumina, carbon fiber carbide, glass fiber, silicon basalt, and biomaterials (e.g., bamboo fibers) have been utilized. Furthermore, with the advancement of technology according to the requirements of applications, manufacturers and researchers are selecting the best possible matrix-reinforcement combination. With the ever-increasing requirement for low weight, high strength, outstanding high-temperature behavior, inertness, and exceptional corrosion resistance materials in automotive and aerospace industries, there has been an encouragement for a progressively emerging activity to fabricate precise composite materials known as aluminum matrix composites (AMCs) [3, 4].

AMCs are high-performance and lightweight materials having the potential to substitute conventional material systems in various advanced appliances. The growing importance of AMCs is a result of material properties such as higher stiffness and strength, advanced compressive strength, lower thermal conductivity, high hardness, enhanced wear resistance, without a noteworthy alteration in ductility, and enhanced stability of dimension. These characteristics enable AMCs as a preferable choice for aerospace and automobile applications. Additionally, AMCs have demonstrated superior mechanical properties at cryogenic as well as higher temperatures, alongside good chemical and corrosion resistance, facilitating it as an outstanding alternative for heavier components [5].

The enhancement in material properties depends upon the adopted processing route, reinforcement material, reinforcement particle size as well as volume fraction. Various fabrication techniques such as centrifugal casting, powder metallurgy, composite casting, stir casting, and liquid infiltration, etc., have been utilized. The AMCs can be reinforced by high carbon steel, ceramic (i.e., Al2O3, SiC, and Y2W3O12, etc.), soft metals (i.e., Cu), and soft non-metals (i.e., fly ash). The variation in mechanical properties with different reinforcement addition is due to reinforcement volume fraction as well as grade of Aluminum matrix. In addition to this, instead of single reinforcement, the combinations of reinforcing materials can also be utilized, wherein both reinforcement complement each other to create hybrid composite materials with advanced performance and features [6, 7].

It is demonstrated from available literature that liquid-phase fabrication is a prominent and viable technique in the development of AMCs with desired shape as well as size, which ultimately influence the mechanical property. The stir casting is one of the liquid state AMCs manufacturing techniques, wherein the matrix phase is aluminum alloy and the reinforcement phase is ceramics. Furthermore, other available liquid phase production methods are compo casting, squeeze casting, and liquid infiltration. In addition to these, the novel liquid phase processing techniques offer the development of net-shaped components having purposefully positioned reinforcement instead of using a common particulate-filled material or preform shape [8, 9].

Although numerous processing techniques are available for the development of AMCs, there are still several issues that need more consideration and research. For example, sometimes problems occur from the improper choice of processing techniques, manufacturing methods, inaccurate collection of process parameters (such as particle size, reinforcement type, melting, and sintering temperatures), and the wrong choice of secondary manufacturing methods. These potential complications can arise while creating defect-free AMCs. This chapter provides a description of various liquid phase processing techniques of AMCs and their relevant properties.

27.2 POTENTIAL REINFORCEMENT MATERIALS FOR AUTOMOTIVE AND AEROSPACE APPLICATION

The reinforcement materials are a very important part of AMCs and responsible for the desirable material properties. The reinforcement material can be categorized into agro-waste ceramic, synthetic ceramic, natural fibers, synthetic fibers, and so on. These reinforcements are described in brief in subsections.

27.2.1 CERAMICS

27.2.1.1 SILICON CARBIDE (SIC)

The SiC incorporated as reinforcement in Al alloy matrix has significantly improved material properties. The incorporation of SiC leads to stopping the creation of aluminum carbide, and results in depreciation of mechanical properties of prepared AMCs. In addition to this, SiC particle size is crucial for the composite integrity because the specific wear rate of AMCs is decreased with SiC particle size increased. Further, the excess SiC volume fraction resulted in the crack formation, increase in friction coefficient and wear rate thus it should be restricted to 15% [10].

27.2.1.2 ALUMINUM OXIDE (ALUMINA)

Aluminum oxide (Al_2O_3) is popular for strong ionic inter-atomic bonding that results in attractive characteristics. It is the stiffest and strongest ceramic oxide, having admirable shape as well as size capabilities and being highly corrosion resistant. Additionally, it has excellent hardness with superior wear resistance. Alumina acquires a higher specific density in comparison to aluminum alloy. The AMCs with alumina reinforcement demonstrate enhanced wettability and remarkably improve compressive strength, breaking strength, ductility, and hardness [11].

27.2.1.3 BORON CARBIDE (B_4C)

AMCs with B_4C as reinforcement exhibited noteworthy wear resistance, tensile strength, good structural rigidity, low density, and high hardness. AMCs with 10 wt.% boron carbide prepared by liquid phase processing showed better strength and hardness compared to SiC reinforced and unreinforced aluminum alloy. The particle size of boron carbide plays an important role in improving material properties, wherein hardness and tensile strength are increased with particle size. Although the addition of boron carbide showed remarkable improvement in certain material properties, its excess volume fraction is detrimental because of agglomeration tendency [12, 13].

27.2.1.4 FLY ASH

The aluminum matrix with fly ash can decrease the cost of preparation. Additionally, fly ash has lower thermal conductivity, lower density, and higher electrical resistivity. These properties are considered beneficial for fabrication of low weight insulator composites. Incorporation of aluminum matrix and fly ash improved the composite integrity by increasing hardness, wear, stiffness, abrasion resistance, and reducing density. In addition to this remarkable improvement in the damping capacity, coefficient of friction, and machinability, is observed. These characteristics are desirable in the automotive industry [14, 15].

27.2.2 AGRO-WASTE-BASED CERAMIC REINFORCEMENTS

There are various agriculture-based residues such as groundnut shell, rice husk, bamboo leaf, sugarcane bagasse, corn cob, coconut shell, periwinkle, and palm kernel which can be adopted as natural ceramic reinforcement in the preparation of AMCs. These agricultural residues/wastes are magnesium and silicon-rich so they can be used as ceramic reinforcement. An AMC prepared by compo-casting using Al alloy and coconut shell ash (CSA) showed lower density, high hardness, tensile, and wear strength due to the presence of MgO and SiO_2. Also, AMC having Al alloy and sugarcane bagasse ash displayed improved mechanical properties, wear, and corrosion resistance [16, 17].

27.3 METHOD OF AMC DEVELOPMENT

27.3.1 STIR CASTING

Stir casting uses a mechanical stirrer for mixing reinforcement into the molten metal. It is an economical and suitable processing technique for the mass production of AMCs. It is also known as the 'vortex technique'. The vortex is formed by the mechanical stirrer for mixing reinforcement to matrix. The setup of stir casting has reinforcement feeder, furnace, and stirrer, as in Figure 27.1. Steps followed in the stir casting are matrix material melting, stirring of molten metal by mechanical stirrer, feeding of reinforcement at a constant rate, stirring of matrix and reinforcement mixture at a constant speed, pouring of the mixture into mold, solidification of mold to form the composite [18].

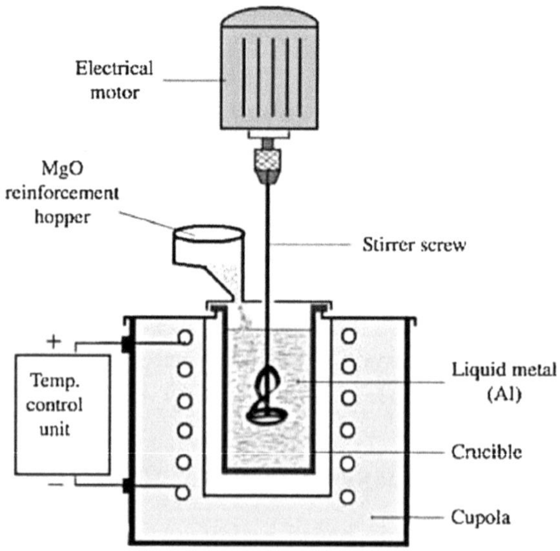

FIGURE 27.1 Stir casting technique for AMCs synthesis.

Source: Reprinted with permission from Ref. [7]. © 2019 The Authors. Published by Elsevier B.V.

The furnace is used to heat and melt the metal matrix. During melting, molten metal may react with the atmosphere and can lead to oxidation of melt. The reinforcement is preheated to remove the volatile substance. The reinforcement is then fed through a feeder in molten metal. The mechanical stirrer is used for mixing reinforcement to the molten metal. The stirrer includes a stirring rod and impeller. The speed of the stirrer is regulated through a variable speed motor. Once the stirring is complete, the pouring of the mixture is carried out through the bottom pouring furnace.

The reinforcement distribution in the matrix governs microstructure and mechanical properties of the composites. Homogeneity in the stir casting depends mainly on stirring parameters. Stirring speed and time, impeller size, position, and blade angle, reinforcement size, and density are dominant factors that influence the reinforcement distribution in the matrix. The uniform distribution of reinforcement can attain optimum mechanical properties, and this can be achieved in stir casting by the optimal choice of stirring parameters [19].

The distribution of the reinforcement particles is affected by stirring speed and time. It was analyzed that microstructure changes along with hardness for different stirring speeds and time. S. B. Prabhu et al. have added SiC reinforcement in A384 alloy and found the optimum stirring speed and stirring time and concludes that higher stirring speed and time give better reinforcement distribution, and thus uniform hardness (110BHN) can be obtained [18]. Furthermore, the effect of blade angle was analyzed and observed that for lower blade angle ($\alpha = 15°$), reinforcement particles disperse down the stirrer. Uniform distribution of particles is observed for blade angle of 30°. While at a higher blade angle ($\alpha > 30°$), particles are concentrated below the tip of the impeller blade. The size of the impeller blade is determined in terms of the furnace diameter [19]. In the case of a smaller blade size, the particles are suspended at the circumference of the crucible. But in case of large blade size, the particles are concentrated near the center of the crucible. The optimal value of the impeller diameter for single stage stirring with a stirring speed of 550 rpm is 0.5 times the diameter of the crucible. Reinforcement is fed to the furnace through the feeder. It is essential to have a continuous flow of particles through the feeder arrangement. A low feed rate (FR) leads to the formation of lumps, and a higher FR results in particle accumulation. Researchers have found that a feed rate of 0.8 g/s to 1.5 g/s is suitable for the proper dispersion of reinforcement particles [20].

27.3.2 INFILTRATION PROCESS

An in-depth look into infiltration processing techniques used to make AMCs is presented in this section. During infiltration, a preform is constructed by reinforcement, then molten metal is poured into the preform and completely fills all open porosities as shown in Figure 27.2 [21].

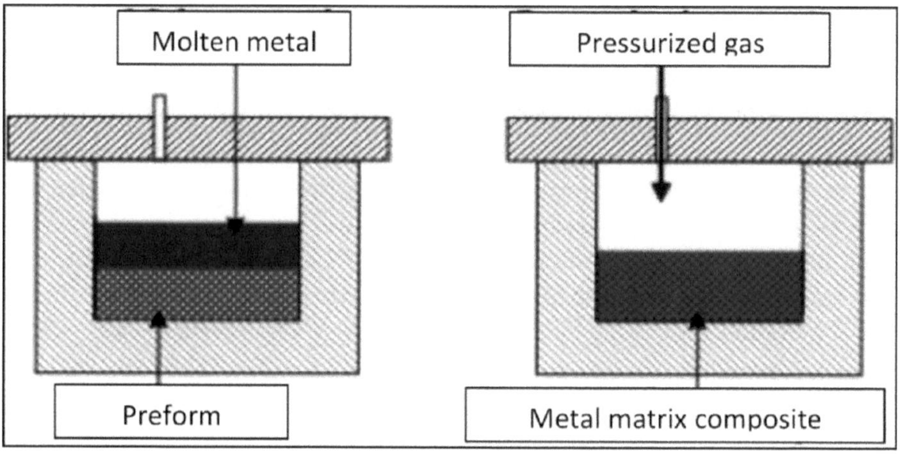

FIGURE 27.2 Infiltration method for development of AMCs.

Source: Adapted from Ref. [7]. © 2019 The Authors. Published by Elsevier B.V.

Flow can be achieved simply by capillarity in environments where molten metal wets the reinforcement. However, mechanical effort must be added in order to overcome the opposing forces caused by drag and capillary formation. The amount of pressure required for mixing reinforcement and matrix depends on the friction impact induced by the molten matrix. Several factors influence how well a liquid alloy wets a ceramic preform, including ceramic preform material, alloy composition, and surface shape. Other important considerations are time and temperature. Numerous researchers have utilized this method for the development of AMCs for application in automotive industries. By utilizing an Al matrix with alumina particles as the reinforcing material, an AMCs was developed and found the particles clustering during processing. The clustering results in certain areas with a greater particle fraction proportion than others. AMC having reinforcement of alumina fiber and alumina particle with a matrix of Al-Cu alloy were fabricated using the lower pressure infiltration method, and the enhancement in wear resistance was observed. Alumina and SiC reinforcement and Al-Mg alloy composite were prepared by pressureless infiltration method and achieved better wettability and penetration in the perform [21].

Spontaneous infiltration has been examined by dipping a pyrolyzed carbon-containing SiC preform into an Al–Si alloy solution in open air. Rapid spontaneous infiltration requires a temperature over 1,400°C, the existence of pyrolyzed carbon, and the presence of silicon carbide particles (SiC) in preforms. They also found that the concentration of Si in the alloy had no effect on infiltration rates, but it had a substantial impact on the microstructures that developed as a result of infiltrating composites [21, 22].

27.3.3 PROCESSING BY IN-SITU (REACTIVE PROCESSING)

There are numerous processes which can be categorized as *in-situ* or reactive processing such as liquid–gas, liquid–solid, and liquid–liquid reactions. Where *in-situ* chemical reactions take place among compounds or elements and elements during synthesis of composite and develop reinforcement phase as shown in Figure 27.3 [23].

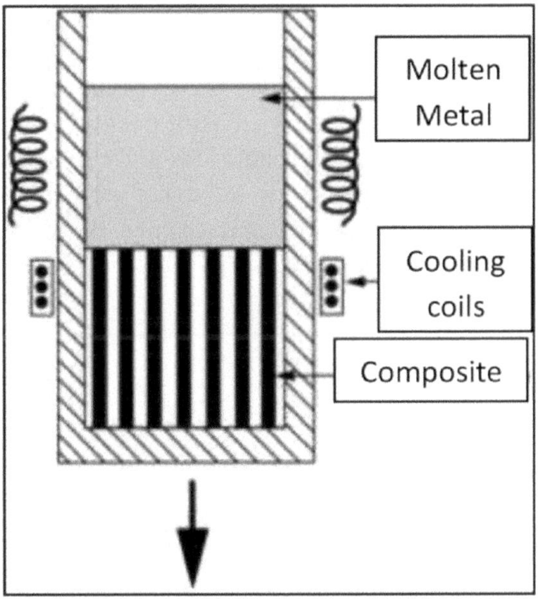

FIGURE 27.3 *In-situ* synthesis of AMCs.

Source: Adapted from Ref. [7]. © 2019 The Authors. Published by Elsevier B.V.

It requires knowledge of reaction kinetics and thermodynamics of processing for the desired product and material properties. The AMCs prepared by *in-situ* process have uniformly distributed finer size reinforcement with strong bonding and a clean interface. Additionally, direct melt reaction, reactive hot pressing, and self-propagating high-temperature synthesis are also based on the *in-situ* principle utilized in the fabrication of AMCs for automotive applications. The direct melt reaction process offers a near net-shape final product at a low cost, thus it is most promising for the preparation of AMCs. *In-situ* direct melt reaction was adopted for development of Al_2O_3/Al composites by chemical reaction of CeO_2 r and Al which occurs spontaneously at 800–900°C temperature. The *in-situ* created Al_2O_3 reinforcement particles are nano-size (100–200 nm). Also, the composites displayed isotropic material characteristics as a result of uniformly distributed Al_2O_3 particles throughout matrix [24].

27.3.4 COMPOCASTING

The processing methodology of compocasting is almost similar to stir casting method but the only difference is the state of matrix material which is semi-solid in this method instead of completely molten state. Thus, sometimes this process is also named as slurry casting process. In this process, the reinforcement material is added to a solidifying melt of matrix material and followed by strong stirring as shown in Figure 27.4. Compocasting offers higher wettability and uniform distribution of reinforcement as compared to stir casting. Al alloy-Fly ash composite and A356-SiC composites were prepared by stir as well as compo casting and found to have enhanced mechanical and microstructure behavior in case of compocasting. Hybrid A356-Al_2O_3-SiC-graphite composites was also successfully prepared by compocasting with enhanced mechanical properties [10].

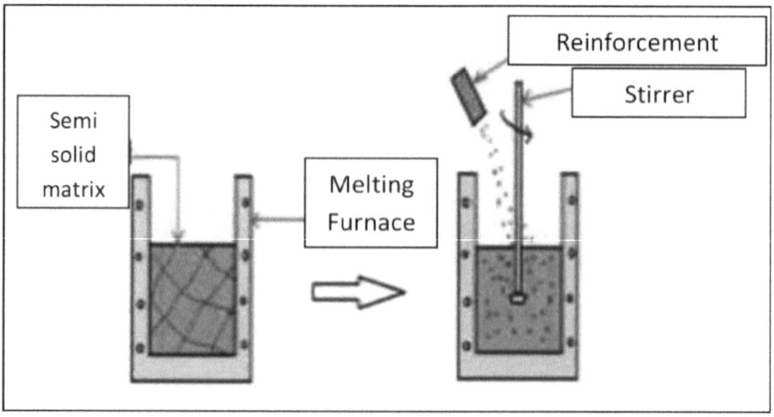

FIGURE 27.4 Experimental setup for compocasting of AMCs.

Source: Reprinted with permission from Ref. [25]. Copyright © 2019 The Society of Manufacturing Engineers. Published by Elsevier Ltd.

27.4 FUTURE DIRECTIONS

This chapter entails the prospective advantages and disadvantages of various liquid phase AMCs preparation routes. Although numerous researchers have investigated secondary treatment to improve material properties of AMCs, but still, it requires more research with more attention over simultaneous application of thermo-mechanical techniques and heat treatment. Additionally, it also requires examining the effect of multi-processing routes (casting, rapid prototyping, powder metallurgy and so on) along with secondary treatment over mechanical, microstructure, and tribological properties of prepared AMCs.

Thus, in order to remove the limitations inherent in the available liquid phase processing methods for AMCs, the following precise proposals are given for future endeavors:

1. Analyze the material properties of AMCs using analytical/finite element modeling techniques to examine the effect of reinforcement type, reinforcement particle size, preparation temperatures, and pressure for chosen matrix-reinforcement combination.
2. Tribological and thermomechanical descriptions of prepared AMCs undergoing secondary treatment by using techniques, i.e., cryo-forging, hot rolling, cryorolling, hot forging, and so on.
3. Investigation of the mechanical, tribological, and microstructural features of AMCs prepared by using multi-synthesis processing.
4. Examination of the effect of applying multiprocessing methods and secondary treatment on material properties.

27.5 SUMMARY AND CONCLUSION

In this chapter, popular manufacturing methods following liquid phase processing are discussed along with their advantages and disadvantages. Several combinations of Al alloy-reinforcement, their performances are discussed with preferred preparation techniques. Additionally, numerous factors that can affect the manufacturing methods chosen are preferred product quality, process performance, applications, and production cost, among others.

1. The stir casting can be utilized for the development of hybrid MMCs. However, this method is successfully used in the preparation

of AMCs with better material performance. It was established that the developed AMCs displayed noticeable enrichment in the mechanical properties having ceramic reinforcement. The inclusion of reinforcement in Al and its alloy resulted in reduction in overall density also and optimum composition gives desired density. Due to these characteristics prepared material found numerous automotive applications. Also, in order to obtain better material properties particle size and stirring is required to be monitored and controlled effectively. Additionally, several mechanical properties such as yield strength, tensile strength, and hardness of AMCs are enhanced with hybrid reinforcement additives. But stir casting process is viable for developing nano-composites due to non-homogeneous nanoparticle distribution with more porosity.

2. In AMCs developed by the infiltration method, a number of porosities with local variation in reinforcement volume fraction is often examined. This technique can be observed as a tremendously versatile and nearby net shape manufacturing method having outstanding ability to control microstructure. However, sometimes local variations of reinforcement particle size and voids are generally observed in AMCs prepared by using this technique.
3. The *in-situ* AMCs manufacturing process creates contamination-free and uniformly dispersed reinforcements, causing better reinforcement-matrix attachment.
4. Spray diffusion turns out to be extremely advantageous for the development of difficult-to-prepare materials.

KEYWORDS

- **agro-waste**
- **aluminum oxide**
- **boron carbide**
- **ceramic reinforcements**
- **compocasting**
- **reactive processing**
- **spray diffusion**

REFERENCES

1. Miracle, D. B., (2005). Metal matrix composites – from science to technological significance. *Compos. Sci. Technol., 65*, 2526–2540. https://doi.org/10.1016/j.compscitech.2005.05.027.
2. Rosso, M., (2006). Ceramic and metal matrix composites: Routes and properties. *J. Mater. Process Technol., 175*, 364–375. https://doi.org/10.1016/j.jmatprotec.2005.04.038.
3. Kaczmar, J. W., Pietrzak, K., &Wlosiński, W., (2000). Production and application of metal matrix composite materials. *J. Mater. Process Technol., 106*, 58–67. https://doi.org/10.1016/S0924-0136(00)00639-7.
4. Bandil, K., Vashisth, H., Kumar, S., Verma, L., Jamwal, A., Kumar, D., et al., (2019). Microstructural, mechanical and corrosion behavior of Al–Si alloy reinforced with SiC metal matrix composite. *J. Compos. Mater., 53*, 4215–4223. https://doi.org/10.1177/0021998319856679.
5. Vinod, B., Ramanathan, S., Ananthi, V., & Selvakumar, N., (2019). Fabrication and characterization of organic and in-organic reinforced A356 aluminum matrix hybrid composite by improved double-stir casting. *Silicon, 11*, 817–829. https://doi.org/10.1007/s12633-018-9881-5.
6. Olszówka-Myalska, A., Szala, J., &Cwajna, J., (2001). Characterization of reinforcement distribution in $Al/(Al_2O_3)_p$ composites obtained from composite powder. *Mater. Charact., 46*, 189–195. https://doi.org/10.1016/S1044-5803(01)00123-1.
7. Garg, P., Jamwal, A., Kumar, D., Sadasivuni, K. K., Hussain, C. M., & Gupta, P., (2019). Advance research progresses in aluminum matrix composites: Manufacturing & applications. *J. Mater. Res. Technol., 8*, 4924–4939. https://doi.org/10.1016/j.jmrt.2019.06.028.
8. Sharma, A. K., Bhandari, R., Aherwar, A., Rimašauskiene, R., &Pinca-Bretotean, C., (2020). A study of advancement in application opportunities of aluminum metal matrix composites. *Mater. Today Proc., 26*, 2419–2424. https://doi.org/10.1016/j.matpr.2020.02.516.
9. Srivatsan, T. S., Ibrahim, I. A., Mohamed, F. A., & Lavernia, E. J., (1991). Processing techniques for particulate-reinforced metal aluminum matrix composites. *J. Mater. Sci., 26*, 5965–5978. https://doi.org/10.1007/BF01113872.
10. Veličković, S., Stojanović, B., Babić, M., Vencl, A., Bobić, I., Bognár, G. V., et al., (2019). Parametric optimization of the aluminum nanocomposites wear rate. *J. Brazilian Soc. Mech.Sci. Eng., 41*. https://doi.org/10.1007/s40430-018-1531-8.
11. Mahesh, K. V., & Venkatesh, C. V., (2018). Effect of ceramic reinforcement on mechanical properties of aluminum matrix composites produced by stir casting process. *Mater. TodayProc., 5*, 2466–2473. https://doi.org/10.1016/j.matpr.2017.11.027.
12. Kummari, N., & Paramesh, M., (2019). Mechanical characterization of aluminum reinforced with boron carbide metal matrix composites. *AIP Conf Proc., 2200*. https://doi.org/10.1063/1.5141172.
13. Shirvanimoghaddam, K., Khayyam, H., Abdizadeh, H., Karbalaei, A. M., Pakseresht, A. H., Ghasali, E., et al., (2016). Boron carbide reinforced aluminum matrix composite: Physical, mechanical characterization and mathematical modelling. *Mater. Sci. Eng. A, 658*, 135–149. https://doi.org/10.1016/j.msea.2016.01.114.
14. Awotunde, M. A., Adegbenjo, A. O., Obadele, B. A., Okoro, M., Shongwe, B. M., &Olubambi, P. A., (2019). Influence of sintering methods on the mechanical properties

of aluminum nanocomposites reinforced with carbonaceous compounds: A review. *J. Mater. Res. Technol., 8*, 2432–2449. https://doi.org/10.1016/j.jmrt.2019.01.026.
15. Ramnath, B. V., Elanchezhian, C., Annamalai, R. M., Aravind, S., Atreya, T. S. A., Vignesh, V., et al., (2014). Aluminum metal matrix composites: A review. *Rev. Adv. Mater Sci., 38*, 55–60.
16. Joseph, O. O., &Babaremu, K. O., (2019). Agricultural waste as a reinforcement particulate for aluminum metal matrix composite (AMMCs): A review. *Fibers, 7*. https://doi.org/10.3390/fib7040033.
17. Alaneme, K. K., Eze, H. I., & Bodunrin, M. O., (2015). Corrosion behavior of groundnut shell ash and silicon carbide hybrid reinforced Al-Mg-Si alloy matrix composites in 3.5% NaCl and 0.3M H 2SO4 solutions. *Leonardo Electron J. Pract. Technol., 14*, 141–158.
18. Prabu, S. B., Karunamoorthy, L., Kathiresan, S., & Mohan, B., (2006). Influence of stirring speed and stirring time on distribution of particles in cast metal matrix composite. J. *Mater. Process Technol., 171*, 268–273. https://doi.org/10.1016/j.jmatprotec.2005.06.071.
19. Sahu, M. K., & Sahu, R. K., (2017). Optimization of stirring parameters using CFD simulations for HAMCs synthesis by stir casting process. *Trans. Indian Inst. Met., 70*, 2563–2570. https://doi.org/10.1007/s12666-017-1119-5.
20. Ravi, K. R., Sreekumar, V. M., Pillai, R. M., Mahato, C., Amaranathan, K. R., Arul, K. R., et al., (2007). Optimization of mixing parameters through a water model for metal matrix composites synthesis. *Mater. Des., 28*, 871–881. https://doi.org/10.1016/j.matdes.2005.10.007.
21. Sree, M. K. M., Ajay, R. L., Rajan, T. P. D., Gupta, M., & Pai, B. C., (2016). Liquid metal infiltration processing of metallic composites: A critical review. *Metall. Mater. Trans. B Process Metall. Mater Process Sci., 47*, 2799–2819. https://doi.org/10.1007/s11663-016-0751-5.
22. Aghajanian, M. K., Rocazella, M. A., Burke, J. T., & Keck, S. D., (1991). The fabrication of metal matrix composites by a pressureless infiltration technique. *J. Mater. Sci., 26*, 447–454. https://doi.org/10.1007/BF00576541.
23. Wang, H., Li, G., Zhao, Y., & Chen, G., (2010). In situ fabrication and microstructure of Al_2O_3 particles reinforced aluminum matrix composites. *Mater. Sci. Eng. A, 527*, 2881–2885. https://doi.org/10.1016/j.msea.2010.01.022.
24. Sahin, Y., &Acilar, M., (2003). Production and properties of SiCp-reinforced aluminum alloy composites. *Compos. Part A Appl. Sci. Manuf., 34*, 709–718. https://doi.org/10.1016/S1359-835X(03)00142-8.
25. Arunachalam, Ramanathan & Krishnan, Pradeep & Rajaraman, Muraliraja. (2019). A review on the production of metal matrix composites through stir casting – Furnace design, properties, challenges, and research opportunities. Journal of Manufacturing Processes. 42. 213-245. 10.1016/j.jmapro.2019.04.017.

INDEX

C

D

Q

R

X

Y

Z